QUAN
COMPUTING
BIBLE 2024

THE DEFINITIVE GUIDE TO MASTERING COMPLEXITY
AND FACE TECHNICAL CHALLENGES OF
TOMORROW'S TRANSFORMATIVE TECHNOLOGY

BY NOAH D. CLEM

© Copyright Noah D. Clem 2024 - All rights reserved.

The content contained within this book may not be reproduced, duplicated, or transmitted without direct written permission from the author or the publisher.

Under no circumstances will any blame or legal responsibility be held against the publisher, or author, for any damages, reparation, or monetary loss due to the information contained within this book. Either directly or indirectly.

Legal Notice:

This book is copyright protected. This book is only for personal use. You cannot amend, distribute, sell, use, quote, or paraphrase any part, or the content within this book, without the consent of the author or publisher.

Disclaimer Notice:

Please note the information contained within this document is for educational and entertainment purposes only. All effort has been executed to present accurate, up-to-date, and reliable, complete information. No warranties of any kind are declared or implied. Readers acknowledge that the author is not engaging in the rendering of legal, financial, medical, or professional advice. The content within this book has been derived from various sources. Please consult a licensed professional before attempting any techniques outlined in this book.

By reading this document, the reader agrees that under no circumstances is the author responsible for any losses, direct or indirect, which are incurred as a result of the use of the information contained within this document, including, but not limited to, — errors, omissions, or inaccuracies.

Table of contents

Introduction 16

Part I: Understanding Quantum Computing 19

CHAPTER 1: INTRODUCTION TO QUANTUM COMPUTING 20

The Evolution from Classical to Quantum Computing **20**
   Classical Computing Fundamentals 21
   Limitations of Classical Computing 22
   The Case for Quantum Computing 22
   The Potential of Quantum Computing 23
   The Qubit Revolution 23
   Quantum Parallelism 24
   The Noisy Intermediate Scale Quantum Era 24
   Quantum Computing Progress 25

Quantum Mechanics and its Relevance to Computing **25**
   Quantum vs. Classical Mechanics 26
   Wave-Particle Duality 26
   Quantum Indeterminacy and Uncertainty 27
   Quantum Tunneling and Teleportation 27
   The No-Cloning Theorem 27
   Quantum Decoherence 29
   The HUP and Quantum Cryptanalysis 29
   Quantum Teleportation 29
   Quantum Algorithms 30

Historical Milestones in Quantum Computing Research — 30
- Quantum Mechanics Beginnings — 30
- Information Theory Origins — 31
- Early Quantum Computing Conjectures — 31
- Post-classical Cryptography — 31
- Quantum Algorithm Breakthroughs — 32
- Early Physical Implementations — 32
- Quantum Error Correction — 32
- Scaling Up Qubits — 33
- Quantum Supremacy — 33
- Noisy Intermediate-Scale Quantum Computing — 33
- Quantum Machine Learning Beginnings — 34

CHAPTER 2: QUANTUM MECHANICS PRIMER — 35

Exploring Quantum Superposition, Entanglement, and Interference — 35
- The double slit experiment and waveparticle Duality — 35
- Quantum superposition and qubits — 37
- Quantum entanglement and non-local — 38
- The uncertainty principle — 38
- Quantum interference and probability amplitudes — 39
- Quantum tunneling — 40
- Quantum decoherence — 41
- Quantum measurements — 41
- Emergence of classicality — 42
- Leggett-garg inequalities — 43
- The Hanbury Brown-Twiss Effect — 44
- Wheeler's Delayed Choice Experiment — 46
- The Quantum Zeno Effect — 46
- Quantum Erasure — 47
- Hidden Variable Theories — 48
- Quantum Non-Demolition Measurements — 48

Mathematical Foundations of Quantum Computing: Dirac Notation and Quantum Gates — 49
- Dirac Notation for Quantum States — 49
- The Bloch Sphere — 50
- Quantum gates and circuits — 51
- The CHSH Game and Bell's Inequality — 51
- Quantizing Classical Systems — 53
- Compatible Observables and Degenerate Codespaces — 53
- Quantum Discord and Information — 54
- Quantum Hardware for Computation and Simulation — 55
- Photonic Quantum Computing — 55

    Topological Quantum Computing   56
    Adiabatic Quantum Computing   56
    Quantum Neural Networks and Machine Learning   57
    Digital vs Analog Quantum Simulation   57
    Decoherence-free subspaces and Noiseless Subsystems   58
    Quantum Money and Cryptocurrencies   58

Quantum Computing as a Paradigm Shift: Implications and Challenges   59
    Exponential Speedups from Quantum Parallelism   59
    Limits Due to Non-deterministic Outputs: 59
    Key Challenges: Decoherence, Noise, and Error Correction: 60
    The Threat to Encryption and Cryptocurrencies: 61
    Concerns around Equity, Ethics, and Misuse: 61

CHAPTER 3: QUANTUM HARDWARE OVERVIEW   63

Quantum Bits (Qubits) and Their Properties   63
    Qubit Definition   64
    Representing Qubits on the Bloch Sphere   64
    Qubit Properties   65
    Qubit Fragility and Decoherence   65
    Readout and Measurement of Qubits   66
    Qubit Connectivity and Control   66
    Qubit Fidelity Metrics   67
    Cost and Scalability of Qubit Technologies   67

Quantum Hardware Architectures: Superconducting Qubits, Ion Traps, and Topological Qubits   68
    Superconducting Qubits   68
    Trapped Ion Qubits   68
    Topological Qubits   70
    Photon Qubits   71
    Quantum Dot Qubits   72
    Hybrid Quantum Systems   73

Quantum Error Correction Techniques: Quantum Codes and Error Mitigation   74
    Sources of Physical Qubit Errors   74
    Quantum Error Correction Principles   74
    Quantum Error Correction Code Examples   75
    Practical Challenges for Realizing Fault Tolerance   78
    Error Mitigation Techniques   78

CHAPTER 4: QUANTUM ALGORITHMS AND QUANTUM SOFTWARE   79

Quantum Algorithm Design Principles: Complexity and Quantum Oracle   80

Leveraging Quantum Parallelism and Superposition	80
Querying Functions via Quantum Oracles	80
Extracting Solutions by Manipulating Superpositions	81
Analyzing Complexity: Query and Time Complexity	81
Notable Quantum Algorithm Speedups	82
Limits to Quantum Speedups	82
Quantum Walk-Based Algorithms	83
Application of Amplitude Amplification	83

In-depth Exploration of Shor's Algorithm for Prime Factorization — 83

Leveraging Quantum Parallelism via Period Finding	84
Applying the Quantum Fourier Transform	84
Combining Quantum and Classical Steps	84
Proving the Quantum Advantage	85
Implementing Modular Exponentiation	85
Analyzing Quantum Speedup Factors	86
Implementation Challenges	86
Implications for RSA Encryption	86
Developing Quantum-Safe Cryptography	87
Practical Implementation Considerations	88
Quantum Error Correction Integrations	88
Resource Optimized Circuits	88
Application Specific Customization	89
Quantum-Classical Hybrid Approaches	89
Analyzing Implementation Tradeoffs	89

Grover's Algorithm: Optimized Search in Unsorted Databases — 90

Quadratic Speedup Over Classical Search	90
Minimizing Query Complexity	90
Optimal Quantum Search	91
Applications of Grover's Algorithm	91
Generalization via Amplitude Amplification	91
Hardware Implementation Considerations	92
Illustrative Examples and Experiments	92
Extensions and Variants	93
Future Research Directions	93

Part II: Mastering Quantum Computing Techniques — 94

CHAPTER 5: BUILDING QUANTUM CIRCUITS — 95

Quantum Circuit Design Principles and Circuit Composition — 95

Qubit Initialization	96
Quantum Gates	97
Gate Rules and Limitations	100
Measurement	100

Reversible Logic	101
Gate Libraries	103
Subcircuits: Enhancing Quantum Circuit Design	105
Mapping to Quantum Hardware	107
Circuit Representation	109
Design Tools	110
Design Methodology	114
Hybrid Quantum-Classical Systems	115
Automated Design	116
Design Verification	118
Design Goals and Tradeoffs	119

Optimizing Quantum Gates: Unitary Operators and Gate Decomposition   119

Unitary Operators	120
Gate Count Reduction	121
Gate Decomposition	123
Commutativity	124
Circuit Depth Reduction	126
Reversible Logic Synthesis	128
Mapping to Available Gates:	130
Ancilla Qubit Reduction: Minimizing Resource Needs	131
Automated Optimization	133
Fidelity Considerations	133
Optimization Challenges: Navigating Complexity	135
Objectives Prioritization	135
Optimization Limits	136
Hardware Co-Design	136
Parameterized Circuits	137

Quantum Circuit Simulation and Verification   137

Quantum State Representation	137
State Propagation	138
Performance Evaluation	138
Debugging	138
Noise Injection	139
Classical vs. Quantum	139
Approximation Techniques	139
Simulator Validation	140
Simulation Libraries	140
Hybrid Simulation	140
Monte Carlo Methods	141
Multiscale Methods	141
Batch Simulation	141
Simulation Performance	142
Continuous Simulation	142

CHAPTER 6: QUANTUM ALGORITHMS IN PRACTICE — 143

Variational Quantum Algorithms for Machine Learning and Optimization — **143**
- Quantum Neural Networks — 144
- Quantum Algorithms for Linear Algebra — 145
- Quantum Combinatorial Optimization — 146
- Implementation and Outlook — 148
- Quantum Circuit Model — 148
- Hybrid Quantum-Classical Optimization — 148
- Training Challenges — 149
- Data Encoding — 149
- Machine Learning Applications — 151
- Combinatorial Optimization — 154
- Quantum Chemistry Simulation — 156
- Quantum Machine Learning for Science — 156
- Hardware Platforms — 156
- Software Stacks — 156

Quantum Approximate Optimization Algorithm (QAOA) for Combinatorial Problems — **157**
- Overview of QAOA — 158
- Applications to Combinatorial Problems — 158
- Practical Implementations — 159
- Hamiltonian Encoding — 159
- Quantum-Classical Hybrid Optimization — 160
- Hardware Implementation — 160
- Software Tools — 163
- Performance Evaluation — 163
- Scaling Up and Limitations — 163
- Use Cases and Applications — 163
- Algorithm Improvements — 164
- Integration with Hybrid Algorithms — 164
- Training and Optimization — 164

Quantum Applications in Chemistry and Materials Science: Simulation and Discovery — **165**
- Quantum Chemistry Simulation — 165
- Quantum Materials Science — 165
- Quantum Machine Learning for Chemical Discovery — 166
- Outlook and Challenges — 166
- Electronic Structure Methods — 167
- Quantum Phase Estimation — 167
- Excited State Algorithms — 167
- Quantum Machine Learning Techniques — 168
- Quantum Error Mitigation — 168
- Time Evolution Methods — 168

Quantum Computational Chemistry Software 168
Algorithm Design and Analysis 169
Molecular Dynamics 169
Quantum Sampling Algorithms 169

CHAPTER 7: PRACTICAL APPLICATIONS OF QUANTUM COMPUTING 170

Quantum Cryptography Protocols: Quantum Key Distribution (QKD) and Beyond **170**
Principles of Quantum Cryptography 171
Quantum Key Distribution (QKD) Implementations 171
Post-Quantum Cryptography 175
Quantum Uncertainty Principle 175
Photon Polarization States 175
BB84 Protocol 176
Commercial QKD Systems 177
QKD Network Implementations 178
Limitations and Challenges 179
Hybrid Security Architectures 179
Improving QKD Range 181
Improving QKD Key Rates 181
Real-World QKD Security 181
QKD Standardization 182
Quantum Computing in Financial Modeling and Portfolio Optimization **182**
Quantum Algorithms for Financial Analysis 183
Portfolio Optimization 183
Financial Risk Analysis and Modeling 183
Quantum Monte Carlo Methods 184
Quantum Machine Learning for Finance 184
Quantum Options Pricing 185
Grover's Algorithm for Portfolio Optimization 185
Quantum Simulations for Risk Analysis 185
Quantum Financial Forecasting 186
Quantum Advantage Validation 186
Quantum Programming Frameworks 187
Quantum Finance Cloud Platforms 187
Quantum Cryptocurrencies 187
Quantum Computing's Role in Supply Chain Management and Logistics **188**
Supply Chain Network Optimization 188
Logistics Optimization 189
Dynamic Replanning and Scheduling 189
Production Planning 189
Inventory Management 190

Demand Forecasting 190
Distribution Optimization 191
Transportation Optimization 191
Quantum Cybersecurity for Supply Chains 191
Supply Chain Simulation 192
Supply Chain Ontologies 192
Operational Integration 192
Adoption Risks 193
Outlook for Commercialization 193

Part III: Navigating Challenges and Future Trends 194

CHAPTER 8: CHALLENGES AND LIMITATIONS 195

Quantum Error Correction: Stabilizer Codes and Fault-Tolerant Quantum Computing **195**
Basic Principles of Quantum Error Correction 196
Sources of Errors in Qubit Systems 196
Qubit Stabilizer Formalism 197
Structure of Stabilizer Codes: 197
Concatenated Quantum Codes: 197
Topological Quantum Error Correction Codes: 198
Classical Error Correction Connections: 198
Efficient Decoding Algorithms: 198
Transversal Logic Gates: 199
Ancilla Qubit Protocols: 199
The Quantum Threshold Theorem 199
Quantum Memory and Repeater Codes 200
Code Transformations 200
Gauge Fixing Protocols 200
Syndrome History Reconstruction 201
Blind Quantum Computation 201
Dynamic Quantum Error Correction 201
Machine Learning for Error Correction 201
Tradeoffs Between Code Overhead and Hardware Quality 202

Scalability Challenges: Quantum Volume and Quantum Supremacy Milestones **203**
Expanding Qubit Arrays: 203
Preserving Qubit Coherence: 203
Minimizing Crosstalk and Control Errors: 204
Benchmarking Quantum Volume: 204
Characterizing Quantum Chaos and Computational Fidelity: 204
Classifying Quantum Error Processes: 205
Realizing Logical Qubits: 205
Embracing Modularity and Heterogeneity: 205

Interfacing Classical and Quantum Systems: 206
Devising Practical Error Correction Strategies: 206
Reducing the Resource Overhead: 206
Designing Quantum Memory Units 207
Realizing a Universal Gate Set 207
Developing Robust Quantum Software Stacks 207
Finding Commercially Valuable Applications 207
Addressing the Cryogenic Challenge 208
Investing in Quantum Engineering 208
Attaining Public and Private Investment 208
Expanding International Collaboration: 209
Setting Realistic Expectations: 209

Overcoming Decoherence: Error Rates and Error Sources 209
Environmental Noise Sources 210
Control Electronics Imperfections 210
Managing Qubit Crosstalk 210
Quantum Measurement Error Processes 211
Manufacturing and Materials Improvements 211
Leveraging Quantum Chaos Studies 211
Sweet Spots and Optimal Bias Points 211
Coherent Pulse Control Techniques 212
Hamiltonian Engineering Approaches 212
Leveraging Quantum Error Correction 212
Specialized Lab Protocols 213
System Design for Noise Isolation 213
Quantum Non-Demolition Qubit Readout 213
Real-Time Feedback and Adaptive Control 213
Characterizing Noise Processes through Models 214
Validating and Updating System Models 214
Designing Robust Pulse Sequences 214
Managing Qubit Variability 215
Leveraging Quantum Error Mitigation 215
Targeting Remaining Decoherence Sources 215

CHAPTER 9: QUANTUM COMPUTING ETHICS AND GOVERNANCE 216

Quantum Information Security: Post-Quantum
Cryptography and Quantum-Safe Solutions **216**
Understanding the Threat Quantum Computers Pose
to Existing Cryptosystems 217
Post-Quantum Cryptographic Standards and Protocols 217
Real-world deployment of Post-Quantum Cryptographic Systems 218
Quantum Key Distribution for Quantum-Secure Communications 218
Transitioning Encrypted Data to Post-Quantum Secure Schemes 219

Ethical Considerations in Quantum Computing: Privacy, Bias, and Fairness — 221
- Privacy Risks and Safeguards in the Era of Quantum Computing — 221
- Recognizing and Mitigating Biases in Quantum Machine Learning — 222
- Trade-Offs Between Utility and Privacy in Quantum Applications — 223
- Fairness in Quantum Cryptanalysis Capabilities and Uses — 224
- Building Public Trust Through Quantum Ethics and Responsible Development — 224
- Responsible Data Practices for Quantum Machine Learning — 225

Regulatory Frameworks and Global Collaboration in Quantum Technologies — 226
- Policymaking for Responsible Quantum Technology Industry Growth — 226
- Multilateral Frameworks for Quantum Technology Cooperation — 227
- Avoiding a Destabilizing Quantum Computing Arms Race — 227
- Quantum Cybersecurity Standards and Best Practices — 228
- Algorithmic Fairness Standards, Testing, and Controls — 229
- Promoting Accessibility and Inclusion in the Quantum Workforce — 229
- Incentivizing Collaborative and Open-Source Quantum Development — 230
- Navigating Trade-Offs in Quantum Regulation — 231
- Incentivizing Responsible Commercial Development — 231
- Promoting Public Engagement and Multidisciplinary Perspectives — 233

CHAPTER 10: FUTURE OF QUANTUM COMPUTING — 233

Quantum Cloud Computing: Quantum-as-a-Service (QaaS) and Cloud Access — 233
- Quantum Computing Cloud Platforms — 233
- Virtual Quantum Labs and Development Environments — 234
- Quantum Computing as a Cloud Service (QaaS) — 235
- Hybrid Quantum-Classical Cloud Workflows — 235
- Edge Quantum Computing — 236
- Cloud-Based Quantum Machine Learning — 236
- Quantum Cloud Security — 237
- Cost Model Evolution — 237
- Cloud Interoperability and Alliances — 237
- Vertical Applications and Partnerships — 238
- Developer Tools and Communities — 238
- Cloud-Based Quantum Workforce Development — 239
- The Outlook for QaaS — 239

Quantum Internet: Building Secure Quantum Networks — 239
- Quantum Teleportation and Cryptography — 240
- Quantum Repeaters: Extending Entanglement Over Long Distances — 240
- Pilot Projects and Testbeds — 240
- Applications: Critical Infrastructure and Financial Systems — 241
- Realizing the Full Quantum Internet — 241

Low-Loss Fiber Networks 242
Satellite-Based Quantum Communication 242
Quantum Network Switching and Routing 242
Quantum Memory Nodes 243
Quantum Network Cybersecurity 243
Integration with Classical Networks 243
Quantum Network Management 244
Deployments in Finance and Government 244
Public Key Quantum Money 244
The Quantum IoT 245
Outlook for Commercialization 245

Speculations on Quantum Computing's Long-term Impact: Industries and Society 245

Disruption of Cryptography and Cybersecurity 246
Financial Services and Blockchain 246
Energy and Materials 247
Healthcare and Drug Development 247
Transportation and Logistics 247
Speculation on Broader Societal Impact 248
Artificial Intelligence 248
Climate Modeling and Science 249
Space Exploration and Aeronautics 249
Intelligence and Defense 249
Food and Agriculture 250
Quantum Workforce Disruption 250
Education Transformation 250
Geopolitical Balance of Power 251
Quantum Legislation and Policy 251
Economic and Social Disruption 251
Reimagining Human-Computer Interaction 252
Rethinking Engineering and Manufacturing 252
Fundamental Discoveries 252
Preparing for a Quantum Society 253

Conclusion 254

Introduction

Quantum computing represents one of the most transformative and disruptive technologies of our time. As we stand at the precipice of the quantum era, this emerging field holds tremendous promise to revolutionize industries, enhance national security, and push the boundaries of our knowledge. Yet, with such enormous potential comes complexity. The peculiarities of quantum physics require new modes of thinking that can be challenging even for seasoned computer scientists to grasp.

It is precisely this challenge that inspires my life's work - to demystify the complexities of quantum computing and make this revolutionary technology accessible to all. I still vividly recall my first quantum computing course as an undergraduate student over two decades ago. While intellectually stimulating, the abstract nature of quantum theory proved difficult to internalize at first. However, through diligent study and hands-on experience programming early quantum devices, I developed an intuition for exploiting quantum mechanical phenomena to enable new forms of information processing. I was hooked, and my passion for this field has only grown stronger since then.

Over the past twenty years, I have dedicated myself to research and education in quantum information science. I have lectured extensively on quantum algorithms and hardware architectures. My publications have explored new techniques in error correction, circuit optimization, and quantum machine learning. I have also advised numerous graduate students who are shaping the future of quantum computing. These experiences have afforded me a broad perspective encompassing both theoretical foundations and practical applications of quantum information processing. It is this holistic viewpoint I wish to share through this book.

The Quantum Computing Bible aims to offer readers a comprehensive yet accessible understanding of this emerging discipline. This undertaking presents unique challenges. Quantum computing intersects many disciplines - physics, mathematics, computer science, and electrical engineering. A basic fluency across these domains is needed to fully appreciate both the fundamentals and applications of quantum computing. Furthermore, this technology raises thought-provoking questions about the nature of information and our ability to model the world. This book will equip readers with the multidisciplinary knowledge to engage meaningfully with such philosophical dimensions.

The book is structured in three parts, taking readers on an enlightening journey from basic concepts to advanced applications. Part I covers the fundamentals of quantum information processing. We begin with essential quantum mechanical principles and discuss their role in enabling quantum speed-ups over classical computing. Critical hardware components such as qubits are explained both conceptually and mathematically. Part II delves into practical techniques for developing quantum algorithms and applications. Readers will learn methods for constructing and simulating quantum circuits as well as implementing celebrated algorithms like Shor's and Grover's. Part III explores the challenges facing quantum computing as it scales to tackle real-world problems. Error correction, decoherence, and ethical concerns are addressed. The book concludes with speculation about the long-term future of the field.

Throughout this journey, the treatment remains accessible to non-experts while providing specialists with mathematical rigor and depth. Approachable analogies illustrate complex topics, from quantum superposition to entanglement. Historical interludes highlight pioneers of quantum computing whose contributions built this field. Chapter summaries and practical exercises reinforce learning objectives. My aim is not only to educate readers but also to inspire the next generation of researchers to propel quantum computing forward.

Make no mistake: mastering quantum computing requires diligence. But the reward is worth the effort. Our growing ability to control quantum systems unlocks capabilities beyond the reach of classical machines. Quantum computing promises to accelerate scientific discovery, improve optimization in ways we cannot yet fathom, and guard information with the strongest encryption allowed by physics. This journey also constantly reminds us that nature still holds surprises - there is so much we have yet to understand. I hope readers will share my sense of wonder at the capabilities quantum phenomena confer.

We stand at a historic moment as quantum computers emerge from research labs into the real world. With competitors racing to develop viable quantum technologies, broad literacy in quantum information science is urgently needed. It is my sincerest hope that The Quantum Computing Bible provides readers with the foundation to engage with quantum computing at all levels, from the curious novice to the experienced professional. Join me on this journey to understand quantum computing, navigate its complexity, and prepare for the quantum future. The most exciting discoveries are still to come!

PART 1

Understanding Quantum Computing

1

CHAPTER 1
INTRODUCTION TO QUANTUM COMPUTING

The Evolution from Classical to Quantum Computing

Classical computing has made significant strides in recent decades, yet it grapples with fundamental limitations. Quantum computing, on the other hand, presents an innovative paradigm by harnessing quantum mechanical phenomena to exponentially enhance computational capabilities. To fully grasp the transformative potential of quantum computing, it is enlightening to juxtapose it with conventional computing and discern the key disparities.

CLASSICAL COMPUTING FUNDAMENTALS

Classical computers rely on transistors and binary digits, known as "bits," to store and process information. These bits exist in one of two states, represented as 0 or 1. All data is encoded in sequences of 0s and 1s, and operations manipulate these bit strings to execute computations. The processing of information in classical computers is linear, one step at a time, due to the exclusive nature of bits. While the miniaturization of transistors has led to exponential growth in processing power, inherent limits persist because of the physical characteristics of classical components.

Modern computers utilize binary digits with two definite states, 0 and 1, to encode and process data. Transistor-based circuits manipulate these binary bit values using Boolean logic gates to perform useful computations. Reducing the size of transistors has facilitated the development of more compact and powerful computers, but heat dissipation concerns hinder further miniaturization. Despite substantial progress, the linear, definitive nature of bit operations in classical computing imposes inherent constraints. Quantum computing seeks to overcome these limits by introducing quantum bits and the concept of superposition.

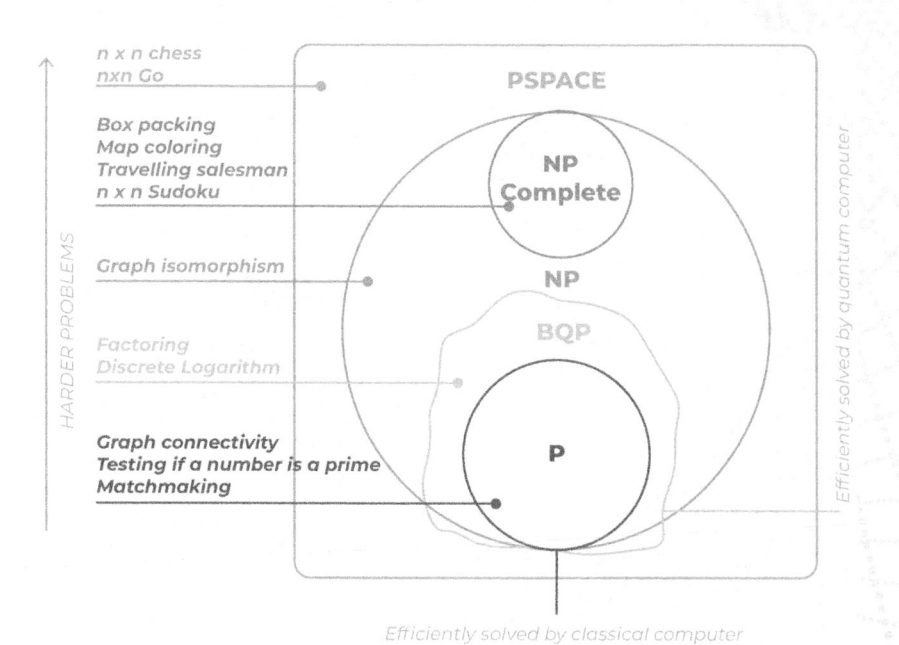

HARDER PROBLEMS

- n x n chess
- nxn Go

- Box packing
- Map coloring
- Travelling salesman
- n x n Sudoku

- Graph isomorphism

- Factoring
- Discrete Logarithm

- Graph connectivity
- Testing if a number is a prime
- Matchmaking

PSPACE
NP Complete
NP
BQP
P

Efficiently solved by quantum computer

Efficiently solved by classical computer

LIMITATIONS OF CLASSICAL COMPUTING

Classical computing is fundamentally restricted by its reliance on definite bits and Boolean logic operations. The evaluation of all possible bit value combinations must occur sequentially, rendering many complex problems intractable. The exponential expansion of Hilbert spaces also imposes constraints on optimization and machine learning applications. These limitations—definiteness, sequence, and scale—serve as the driving force behind the quantum computing paradigm.

Definiteness in classical bits restricts their information representation and manipulation capabilities. Boolean logic gates and sequences of 0 and 1 values enforce linearity in computational processes. The exponential scale of numerous real-world problems further strains classical computing capacity, as evident in challenges faced during machine learning training with massive datasets. While Moore's Law has undeniably improved classical computing, quantum mechanics introduces innovative approaches to surmount these inherent limitations.

THE CASE FOR QUANTUM COMPUTING

Quantum computing aims to revolutionize computing by leveraging quantum mechanical phenomena. Quantum bits or "qubits" can exist in superposition, enabling massively parallel information representation and processing. Entanglement generates correlations, allowing quantum algorithms to solve problems intractable for classical computers rapidly. Quantum computing heralds immense untapped computational potential.

Qubits can represent information as superpositions of 0 and 1, enabling simultaneous evaluation of multiple states. Quantum parallelism allows quantum computers to process immense combinations exponentially faster than classical serial processing. Entanglement generates non-classical correlations between qubits, permitting optimization and machine learning algorithms to rapidly converge on quality solutions within enormous possibility spaces. Quantum computing can transcend the scaling challenges facing classical computing.

THE POTENTIAL OF QUANTUM COMPUTING

Quantum computing promises solutions for diverse, complex problems across fields, including cryptography, optimization, machine learning, and materials science. Applications range from financial portfolio optimization to chemical simulations. Future quantum networks could provide unconditionally secure communication. The potential impact to society is vast as quantum capabilities exceed classical computing constraints.

Quantum algorithms like Shor's factor large integers exponentially faster, breaking current encryption schemes. Optimization solvers harness quantum parallelism for logistics, scheduling, and financial portfolio problems with vast combinations. Quantum machine learning trains complex models like deep neural networks rapidly using quantum-enhanced training algorithms. Quantum simulation accurately models atomic and subatomic interactions for chemistry and materials science. The expansive capacity of quantum computing creates disruptive potential across information sciences.

THE QUBIT REVOLUTION

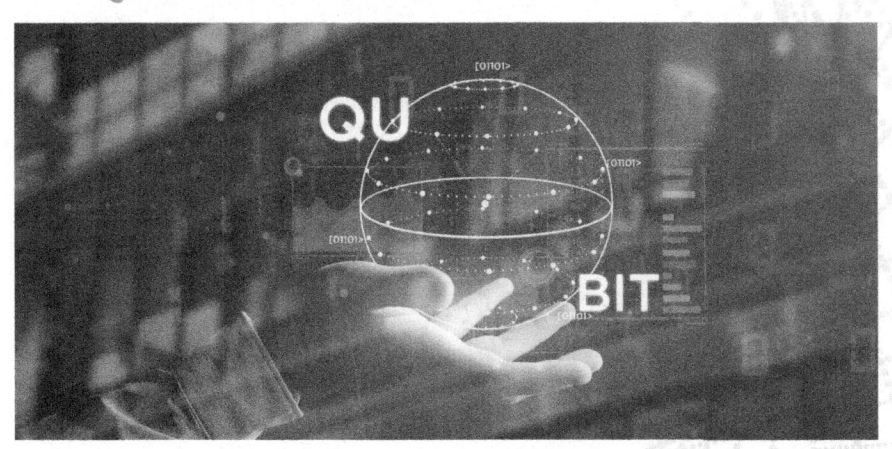

The quantum bit, or qubit, stands as a revolutionary departure from the classical bit. Unlike its classical counterpart, a qubit is not confined to a definitive 0 or 1; it can exist in a superposition of both states simultaneously. This unique characteristic enables a massively parallel representation of information, leading to an exponential increase in computational capability relative to classical bits. The qubit serves as the fundamental unit that empowers quantum computers to transcend traditional limits.

Unlike classical bits, qubits can represent 0 and 1 simultaneously through quantum superposition. This allows for the parallel evaluation of all possible combinations of qubit states, providing an exponential information advantage over classical bits. Additionally, qubits exhibit entanglement, giving rise to non-classical correlations. The counterintuitive nature of qubits enables quantum algorithms to operate within Hilbert spaces that are beyond the reach of classical computing. The qubit is the cornerstone of quantum computing's extraordinary power.

QUANTUM PARALLELISM

Quantum parallelism leverages qubit superposition to evaluate all possible solutions simultaneously. While this does not provide the final output, it creates an interference pattern, enabling algorithms to rapidly hone in on quality solutions. This provides exponential speedup over classical serial processing and underpins many quantum applications.

Qubit superposition represents all 0 and 1 permutations in parallel. While measurement collapses, the interference of amplitudes creates a pattern algorithm exploit to increase computational speed exponentially. This quantum parallelism enables everything from database search to machine learning optimization far faster than is classically possible. It underscores why quantum is revolutionary.

THE NOISY INTERMEDIATE SCALE QUANTUM ERA

While fully fault-tolerant quantum computers remain a long-term aspiration, the current landscape is marked by rapid progress in the noisy intermediate-scale quantum (NISQ) era. Present-day quantum processors, housing tens to hundreds of noisy qubits, are already showcasing quantum advantages. Continuous improvements in algorithms and hardware will further elevate NISQ capabilities.

Current quantum computers house approximately 100 imperfect qubits susceptible to noise. Although error-corrected qubits pose challenges, these noisy intermediate-scale quantum processors demonstrate computational superiority over classical supercomputers in specific applications. Ongoing advancements in quantum error correction, fault tolerance, and qubit quality promise to broaden NISQ capabilities and propel the evolution of quantum computing.

QUANTUM COMPUTING PROGRESS

The past decades have witnessed enormous leaps in translating theoretical quantum computing into reality. Feynman, Deutsch, and other pioneers established the conceptual foundation. Hardware, materials science, and algorithm innovations at companies including Google, IBM, and D-wave have rapidly accelerated experimental quantum computing.

Early thought experiments envisioning quantum computational models gave rise to formal theories and principles. Hardware implementations were stubborn, but steady materials science advances in manufacturing and controlling qubits enabled practical demonstrations. With companies investing heavily, and rapid algorithmic innovations like hybrid quantum-classical schemes, experimental quantum computing has progressed remarkably over recent years.

Quantum Mechanics and its Relevance to Computing

Quantum mechanics describes the behavior of nature at subatomic scales and introduces counterintuitive concepts fundamental to quantum computing. Phenomena including superposition, entanglement, and uncertainty enable capabilities to surpass classical physics limits. An overview of these relevant quantum principles provides key context.

QUANTUM VS. CLASSICAL MECHANICS

Classical physics accurately describes our macroscopic world. However, examining nature at smaller scales reveals limitations. Quantum mechanics radically departs from classical intuition, yet its predictions align with experimental observations. This departure is the source of quantum computing's power.

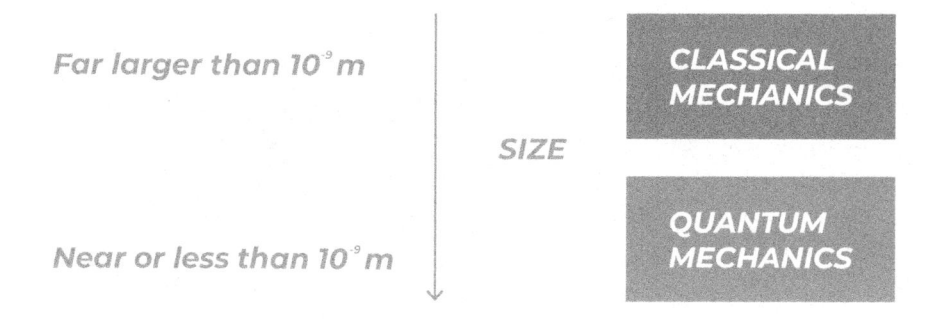

Far larger than 10^{-9} m — **CLASSICAL MECHANICS**

SIZE

Near or less than 10^{-9} m — **QUANTUM MECHANICS**

Isaac Newton's laws precisely describe observable classical mechanics, but they fail at subatomic levels where quantum effects dominate. While counterintuitive, quantum theory makes extraordinarily accurate predictions confirmed experimentally. By harnessing quantum phenomena lacking classical analogs, quantum computers open new computing frontiers.

WAVE-PARTICLE DUALITY

Quantum entities exhibit properties of both particles and waves. This enables superposition, where matter essentially exists in multiple potential states simultaneously. Superposition is foundational for enabling massive parallelism in quantum computers.

Matter fundamentally has wave-like properties at small scales according to quantum mechanics. This differs radically from classical particles occupying definite positions. Wavelike superposition allows the representation of 0 and 1 qubit states in parallel. This property enables exponential scale-up compared to classical bits

QUANTUM INDETERMINACY AND UNCERTAINTY

Quantum measurements introduce an intrinsic randomness absent in classical systems. The Heisenberg Uncertainty Principle implies that fixed values cannot describe quantum states. This uncertainty is harnessed in quantum applications like cryptography.

In classical physics, properties like position and momentum are fixed values permitting deterministic predictions. Quantum theory reveals nature is inherently uncertain. Measurements of conjugate variables like position and momentum cannot be simultaneously determined. This uncertainty enables quantum cryptography and poses challenges for error correction.

QUANTUM TUNNELING AND TELEPORTATION

Quantum tunneling allows particles to exhibit behaviors disallowed in classical systems, like traversing energy barriers. Quantum teleportation transports qubit states across entanglement links through nonlocal correlations. These phenomena have computational utility.

Classically, objects cannot penetrate energy barriers without sufficient kinetic energy. However, quantum tunneling permits this due to the probabilistic spread of matter waves. In addition, quantum entanglement enables the teleportation of qubit states over channels using nonlocal correlations. These effects are harnessed in quantum algorithms and hardware..

THE NO-CLONING THEOREM

The no-cloning theorem is a fundamental principle in quantum mechanics that states that it is impossible to create an identical copy of an arbitrary unknown quantum state. This theorem has profound implications for quantum information processing, particularly in the context of quantum key distribution (QKD) and quantum cryptography.

The theorem was first proposed by Wootters and Zurek, and independently by Dieks, in 1982. It is based on the linearity and unitarity of quantum mechanics, which are essential for preserving the probabilistic nature of quantum measurements and the coherence of quantum states.

To understand the no-cloning theorem, let's consider a quantum state $|\psi\rangle$ that we want to clone. A hypothetical cloning machine would take this state as input and produce two identical copies of the state as output:

$$|\psi\rangle \rightarrow |\psi\rangle \otimes |\psi\rangle$$

However, the no-cloning theorem proves that such a machine cannot exist for arbitrary unknown quantum states.

Proof of the No-Cloning Theorem:

Suppose we have two arbitrary quantum states $|\psi\rangle$ and $|\varphi\rangle$, and a hypothetical cloning machine U that can clone these states:

$$U(|\psi\rangle \otimes |0\rangle) = |\psi\rangle \otimes |\psi\rangle$$
$$U(|\varphi\rangle \otimes |0\rangle) = |\varphi\rangle \otimes |\varphi\rangle$$

Where $|0\rangle$ represents the initial state of the cloning machine.

Now, consider the case where we want to clone a superposition state $|\psi\rangle + |\varphi\rangle$:

$$U((|\psi\rangle + |\varphi\rangle) \otimes |0\rangle) = (|\psi\rangle + |\varphi\rangle) \otimes (|\psi\rangle + |\varphi\rangle)$$

However, due to the linearity of quantum mechanics, the cloning machine must act on the superposition state as follows:

$$U((|\psi\rangle + |\varphi\rangle) \otimes |0\rangle) = U(|\psi\rangle \otimes |0\rangle) + U(|\varphi\rangle \otimes |0\rangle)$$
$$= |\psi\rangle \otimes |\psi\rangle + |\varphi\rangle \otimes |\varphi\rangle$$

This result is different from the desired output $(|\psi\rangle + |\varphi\rangle) \otimes (|\psi\rangle + |\varphi\rangle)$, which shows that the cloning machine cannot clone arbitrary superposition states.

QUANTUM DECOHERENCE

The fragility of quantum superpositions poses challenges. Environmental interactions induce decoherence, collapsing qubit superpositions into classical states. Quantum noise remains a key hurdle for practical error-corrected quantum computing.

Qubits can easily lose their quantum properties through interaction with the external environment. This quantum decoherence reverts superpositions to classical states. Engineering quantum systems to minimize noise and maintain coherence remains a massive challenge. Error-corrected logical qubits will be needed to create stable large-scale quantum computers.

THE HUP AND QUANTUM CRYPTANALYSIS

Quantum cryptanalysis leverages the Heisenberg Uncertainty Principle to defeat classical ciphers like RSA. However, quantum key distribution protocols can enable unconditionally secure communication between partners sharing entangled particle pairs.

Classical public key cryptography relies on hard mathematical problems like factoring large numbers. However, Shor's quantum algorithm exploits uncertainty to efficiently factor hitherto secure keys. This jeopardizes classical encryption but also enables unbreakable quantum key distribution, harnessing shared entanglement.

QUANTUM TELEPORTATION

Quantum teleportation demonstrates the bizarre nature of entanglement, allowing qubit-state information to be reconstructed over quantum channels faster than light could travel. This has implications for quantum communication and networking.

By leveraging shared entanglement and classical communication, quantum teleportation can reliably transport qubit states between distant locations. This "spooky action at a distance" highlights the non-local nature of quantum information. Networking remote quantum computers will require such teleportation techniques.

QUANTUM ALGORITHMS

Quantum algorithms like Shor's and Grover's leverage superposition, entanglement, and uncertainty to achieve exponential speedup over classical counterparts for factoring, search, simulation, and optimization. This demonstrates quantum computing's immense potential. Shor's algorithm can factor large numbers exponentially faster than classical algorithms, with major cryptanalysis implications.

Grover's search achieves quadratic speedup over classical search, enabled by superposition and quantum parallelism. Quantum simulation efficiently models complex molecular and physical systems that are intractable classically. These demonstrate the origins and immense power of quantum speedup.

Historical Milestones in Quantum Computing Research

The conceptual origins of quantum computing stem back almost a century to pioneering work blending quantum physics and information theory. Through subsequent decades, steady theoretical and technological progress built the foundations for today's quantum information revolution. Reviewing key historical milestones provides essential context.

QUANTUM MECHANICS BEGINNINGS

Early 20th-century pioneers like Planck, Einstein, Bohr, and Schrodinger established quantum physics, revealing a radically counterintuitive description of nature at microscopic scales, and laying the groundwork for quantum computing possibilities.

Max Planck's introduction of energy quantization catalyzed the origins of quantum physics at the turn of the 20th century. Einstein contributed concepts like photons and wave-particle duality. Niels Bohr developed a quantum model for hydrogen's atomic structure. Erwin Schrodinger's equation described matter waves. Together, these pioneers established the strange rules of quantum mechanics underpinning quantum information processing.

INFORMATION THEORY ORIGINS

Information theory emerged alongside quantum theory in the early 20th century, driven by pioneers like Nyquist, Hartley, von Neumann, and Shannon. This blending of information and physics concepts would prove integral to quantum computing.

Harry Nyquist and Ralph Hartley developed early frameworks characterizing communication channel capacity. John von Neumann contributed mathematical rigor to quantum theory and collaborated with information theorists. Claude Shannon founded modern information theory quantifying communication limits. Together, these works interfaced with information and physics, foreshadowing quantum information processing.

EARLY QUANTUM COMPUTING CONJECTURES

Visionary physicists such as Feynman and Deutsch recognized the intrinsic connections between quantum mechanics and computation. Their thought experiments paved the way for radical improvements in computational power if quantum effects could be effectively harnessed.

In 1982, Richard Feynman proposed that quantum systems could efficiently simulate physical systems beyond the reach of classical computers. Building on this, David Deutsch theorized in 1985 that a quantum Turing machine could gain a substantial advantage by exploiting superposition. These pioneering thought experiments laid the conceptual foundations for quantum computational complexity theory.

POST-CLASSICAL CRYPTOGRAPHY

Quantum computing jeopardizes classical public key cryptography. However, Wiesner's early work introduced quantum protocols enabling unbreakable cryptographic key exchange based on Heisenberg uncertainty and entanglement.

Stephen Wiesner's prescient 1969 paper proposed unbreakable quantum money and communications exploiting Heisenberg uncertainty. This pioneered quantum cryptography protocols shielding secrets from adversaries by encoding information in quantum states. Charles Bennett and Gilles Brassard later formalized quantum key distribution schemes.

QUANTUM ALGORITHM BREAKTHROUGHS

Inspired by Deutsch's insights, groundbreaking algorithms like Shor's for factoring and Grover's quantum search showcased the potential for quantum computers to revolutionize computing through exponential speedups over classical counterparts.

In 1994, Peter Shor conceived the quantum factoring algorithm, demonstrating significant speedup potential compared to classical factoring approaches. Lov Grover's 1996 quantum search algorithm offered quadratic speedup over classical search through quantum parallelism. These breakthroughs underscored the immense latent capability of quantum computers.

EARLY PHYSICAL IMPLEMENTATIONS

Laborious physical implementations of few-qubit processors in the 1990s and 2000s demonstrated basic quantum effects while underscoring challenges around controllability, scalability, and error correction necessary to build capable quantum computers.

Following Shor and Grover's algorithms, experimental physicists constructed rudimentary quantum processors using systems like ion traps and superconducting loops to represent quantum information. While limited to just a few noisy qubits, these provided a beachhead toward scalable physical realizations of more capable quantum computers.

QUANTUM ERROR CORRECTION

The introduction of quantum error correction codes addressed decoherence challenges, mitigating the threat to quantum superpositions. This breakthrough indicated that overcoming physical fault tolerance constraints in quantum computing was indeed feasible.

As quantum systems interacted with their environments, qubits faced rapid decoherence into classical states. The discovery of quantum error correction codes, utilizing redundant qubits to detect and correct errors, illuminated a path forward for reliable quantum computing despite the presence of noise and imperfect control

SCALING UP QUBITS

Steady materials science and fabrication improvements have enabled rapid qubit number scaling on experimental quantum processors over the past decade. Today's noisy intermediate-scale quantum era with 50-100 qubits is advancing toward the 1000s of qubits needed for fault tolerance.

Increasing qubit counts while maintaining coherence remained challenging through the 2000s. However, material advances and refined controls permitted organizations like Google, IBM, and Rigetti to double qubit numbers yearly. While symptomatic noise persists, progress toward the hundreds and thousands of physical qubits needed for error-corrected logical qubits continues.

QUANTUM SUPREMACY

Google's 2019 Sycamore demonstration showed a 53 qubit processor could perform a narrowly defined computation in 200 seconds (about 3 and a half minutes) that would take a supercomputer 10,000 years, achieving a major quantum computing milestone.

In late 2019, Google announced its 53 qubit Sycamore processor had performed a randomly sampled quantum circuit sampling problem outside the reach of the Summit supercomputer in a reasonable time frame. While limited in scope, this milestone "quantum supremacy" demonstration proved the potential of quantum computers to solve problems intractable for classical systems.

NOISY INTERMEDIATE-SCALE QUANTUM COMPUTING

The era of noisy intermediate-scale quantum computing is seeing today's 50-100 qubit devices deliver quantum advantage through hybrid algorithms resilient to noise and errors. Near-term advances will expand these capabilities.

While universal error-corrected quantum computers remain longer-term goals, present NISQ processors with approximately 100 imperfect qubits are demonstrating quantum advantage. Hybrid quantum-classical algorithms and error mitigation techniques are enabling meaningful computations on noisy hardware likely to deliver commercial value and advance the field.

QUANTUM MACHINE LEARNING BEGINNINGS

Quantum techniques offer potential speed and modeling advantages for machine learning tasks, including dimensionality reduction, optimization, and neural network training. Early quantum machine learning algorithms highlight significant promise.

Classical machine learning often needs help with high-dimensional data. Quantum techniques like quantum principal component analysis can exponentially reduce dimensions. Quantum-enhanced optimization and sampling have accelerated training in limited demonstrations. Further developing quantum machine learning will be critical to capitalize on quantum advantage.

CHAPTER 2
QUANTUM MECHANICS PRIMER

Exploring Quantum Superposition, Entanglement, and Interference

THE DOUBLE SLIT EXPERIMENT AND WAVE-PARTICLE DUALITY

The double slit experiment beautifully demonstrates the fundamental mystery of quantum mechanics. When particles like electrons or photons are fired at a barrier with two narrow slits, an interference pattern emerges on the screen behind it. This pattern with alternating bright and dark bands would be expected for waves splitting and recombining. However, we are firing discrete particles. So, how can individual particles interfere with each other as if they are waves? This points to a counterintuitive quality of the quantum world - every particle can also act as a wave. This is called wave-particle duality. The probability waves passing through both slits interfere to determine where the particle may be detected. This experiment established that the physics of tiny atomic particles is radically different from the classical physics governing everyday objects.

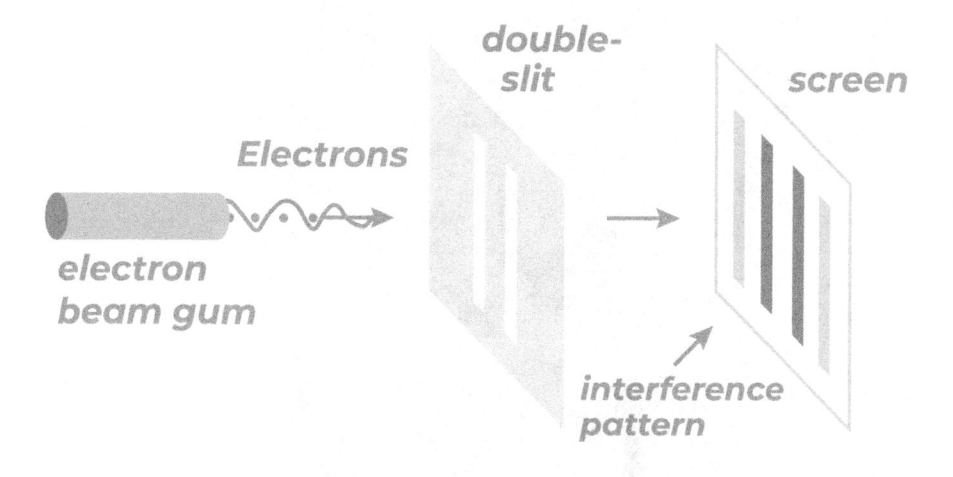

Electrons

electron beam gum

double-slit

screen

interference pattern

The double slit experiment was originally carried out in the early 1800s by scientist Thomas Young to demonstrate the wave nature of light. He observed the interference pattern when shining light through two closely spaced slits. But in modern versions, the same effect is seen even when sending photons, electrons, or other particles through the slits one at a time. The buildup of an interference pattern, even with individual quanta, conclusively showed that the particle-like photons and electrons possessed an intrinsic wave-like nature. This challenged the notion that light and matter existed either exclusively as particles or waves. By uncovering wave-particle duality, the double slit experiment contradicted classical intuitions and helped usher in a radically new quantum understanding of nature's fundamental constituents.

QUANTUM SUPERPOSITION AND QUBITS

This brings us to the concept of superposition. According to quantum theory, particles do not exist in a definite state but as a superposition or combination of multiple possible states. For example, an electron can be in a superposition of two energy levels or spin states at the same time. The probability for measuring each state upon observation can be specified as complex numbers using the Dirac notation. This ability of quantum systems to exist in multiple states allows for massive parallelism when applied to computation. Qubits are quantum bits that can represent a 0, 1, or a superposition of both. Just like classical bits form the basic units of classical computers, qubits form the basic information units of quantum computers. A system of multiple qubits can exist in all permutations of 0s and 1s simultaneously. As we will see later, this enables certain computations to be done in parallel, leading to exponential speedups.

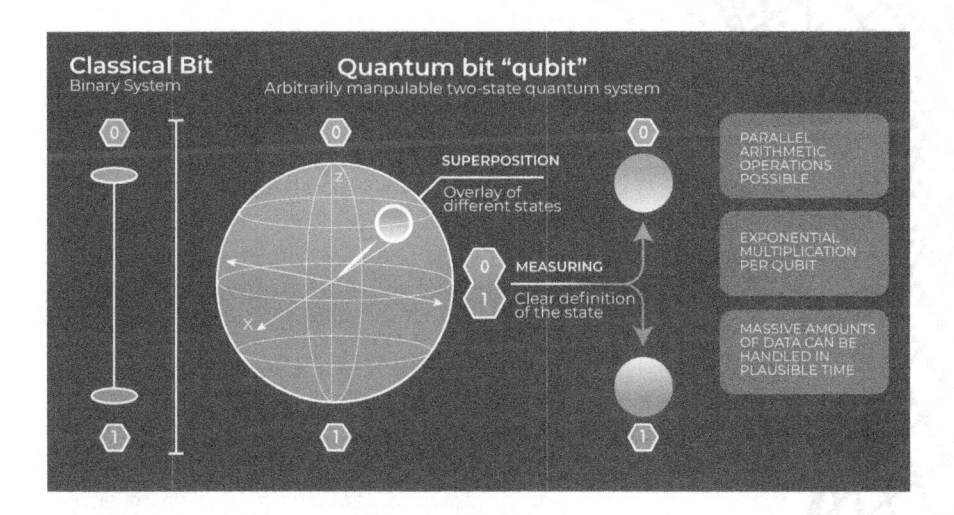

Superposition is closely tied to the wave nature of matter seen in the double-slit experiment. Waves can be added together or superimposed constructively or destructively to form new combined waves. Similarly, quantum states can be superimposed to form more complex quantum systems. Consider a simple qubit. Instead of classically being limited to states 0 or 1, the qubit can exist in any superposition $α|0> + β|1>$ where $α$ and $β$ are probability amplitudes. Measurement will randomly collapse this to 0 with probability $|α|2$ or to 1 with probability $|β|2$ in line with quantum uncertainty. Qubits in superpositions of 0 and 1 can exploit massive parallelism during computation. A quantum computer with 500 qubits occupies 2500 states simultaneously - more than the number of atoms in the observable universe! This exponential scale-up underlies the tremendous power of quantum computation.

QUANTUM ENTANGLEMENT AND NON-LOCALITY

Another quintessential quantum effect is entanglement. When two particles interact, they can become entangled, and their properties then become correlated in a way that transcends the usual laws of space and time. Measuring one particle instantaneously determines the state of the other even if they are light years apart. This leads to what Einstein called "spooky action at a distance." Entangled particle pairs can be used to enable two distant parties to exchange information in ways not possible classically. This can be harnessed for cryptographic protocols like quantum key distribution that could enable ultra-secure communication networks. Entanglement also enables quantum teleportation and forms the basis for quantum error correction schemes.

Entanglement was termed by Einstein as "spukhafte Fernwirkung", or spooky action at a distance because it seemed to violate special relativity. If information transfer is limited by the speed of light, how can the measurement of one particle instantaneously affect the other? Einstein believed quantum mechanics must, therefore, be incomplete. However, John Bell formalized this concept mathematically, showing that entangled particles really exhibit stronger than classical correlations. Experiments to test Bell's inequalities have conclusively demonstrated that entanglement is physically real. Entanglement forces us to accept that at the quantum scale, physical systems exhibit a holistic behavior not explainable by classical notions of cause and effect. Understanding and harnessing entanglement is key to quantum information processing.

THE UNCERTAINTY PRINCIPLE

The uncertainty principle is another pillar of quantum theory that limits our ability to simultaneously measure complementary properties like position and momentum. The act of observation disturbs the system, so we cannot know both quantities beyond a minimal level of uncertainty. This prevents us from precisely predicting the future state of quantum systems. Instead, we can only calculate the probability of different outcomes upon measurement. The fuzziness introduced by the uncertainty principle is what accounts for many of the peculiar aspects of the quantum world. But, the probabilistic nature of quantum states also allows them to encode information in new ways.

The uncertainty principle was formulated by Werner Heisenberg in 1927. It states that for complementary variables like position and momentum, their uncertainties in measurement satisfy $\Delta x \Delta p \geq h/4\pi$ where h is Planck's constant. There is a fundamental tradeoff - the more precisely we know one variable, the less precisely we can know the other. At the quantum scale, the act of measurement disturbs the system enough that we cannot attain certainty in both quantities. This places fundamental limits on what we can know about a quantum system's state at any given time. The uncertainty principle leads to several quintessential quantum effects and provides intrinsic protection for quantum information. The information encoded across complementary degrees of freedom cannot be accessed without undergoing some disturbance.

QUANTUM INTERFERENCE AND PROBABILITY AMPLITUDES

Now, let's delve back into the phenomenon of interference observed in the double-slit experiment. How does interference manifest at the quantum level? According to quantum mechanics, particles traverse all conceivable paths from the source to the detector. Each path is associated with a complex number termed the probability amplitude. The likelihood of detecting the particle at a specific location is governed by the interference among probability amplitudes from all paths. Constructive interference arises when waves combine to produce maximum intensity, enhancing the probability of detection. Conversely, destructive interference leads to cancellation, resulting in a lower probability. This wave-like interference of probability amplitudes is fundamental to numerous quantum algorithms.

Probability amplitudes, complex numbers representing the likelihood of a quantum system existing in a particular configuration, can be added following the laws of vector addition. The probability is then proportional to the squared magnitude of the resulting amplitude. This mechanism allows probabilities from multiple configurations to interfere like waves—either constructively adding up or destructively canceling out. In the double-slit experiment, positive interference between amplitudes from both slits creates bright fringes at certain locations on the screen, while destructive interference reduces the probability at other locations. Calculating these interference effects among an exponential number of paths enables many quantum algorithms to outperform classical approaches.

QUANTUM TUNNELING

Quantum tunneling is an effect with no classical analog, allowing particles to traverse a barrier even when lacking sufficient classical energy to overcome it. This phenomenon arises from the wave nature of particles and their probabilistic distribution over space. Quantum tunneling facilitates vital physical processes like nuclear fusion in the sun, electron microscope scans, and flash memory storage. It also forms the basis for quantum annealing algorithms. However, tunneling can lead to errors in quantum computers, as we'll discuss later.

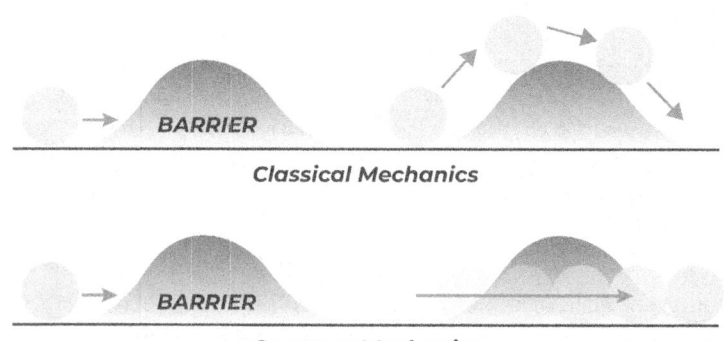

Classical Mechanics

QuantumMechanics

Quantum tunneling exploits the wave property of matter, enabling waves to exhibit behaviors such as diffraction, allowing them to propagate to classically forbidden regions. Even if a particle lacks the classical energy to traverse a barrier, its wavefunction penetrates the barrier with some non-zero probability, resulting in tunneling to the other side. In quantum computing, tunneling allows qubits to transition into lower energy states by bypassing energy barriers. This is utilized in adiabatic quantum optimization and quantum annealing. However, tunneling can inadvertently flip qubit states, causing computational errors. To prevent unwanted tunneling events, careful qubit design and error correction techniques are employed in gate-model quantum computers.

QUANTUM DECOHERENCE

A key challenge is that quantum superpositions are extremely fragile. Any slight interaction with the environment collapses the superposition to a single state, an effect called quantum decoherence or dephasing. This can lead to errors and limits the time over which quantum algorithms can run. Quantum decoherence arises from the non-classical nature of quantum superpositions. They appear to "leak" or dissipate into the wider environment. Minimizing decoherence is critical for building practical quantum computers. Techniques like quantum error correction, noise-resistant encodings, and isolation from the environment help mitigate decoherence and will be explored later.

Quantum superposition and entanglement with external degrees of freedom lead to decoherence. This destroys the phase relationships between basis states, making the system classical. To prevent it, quantum computers operate at extremely low temperatures, and clever qubit designs "insulate" against noise. Still, some decoherence is inevitable, so error correction is needed. The threshold theorem states reliable quantum computing is possible if the error rate is below a threshold. Much work remains to build quantum computers crossing this error threshold. Overcoming decoherence will allow us to fully harness the powers of superposition and entanglement.

QUANTUM MEASUREMENTS

Quantum measurements play a pivotal role in extracting information from qubits after computation. However, the act of measurement irreversibly collapses the quantum state to an observable basis state, annihilating superpositions. The squared amplitude gives the probability of observing a specific state. Partial measurements can also be conducted by coupling ancilla qubits and then measuring those without directly observing the system qubits. Weak measurements only partially collapse the wavefunction. Quantum circuits must be designed to incorporate optimal measurement routines for reading out results.

Careful timing of measurements and utilizing ancilla measurement qubits prevent decoherence of the main computing qubits. Some algorithms require intermediate weak or partial measurements. Noise and imperfect measurements can yield incorrect results, so quantum error correction is employed. Advances like bosonic qubits with inherently robust readout states, efficient photon number detectors, and minimally disruptive ancilla schemes continue to enhance the quantum measurement process, which is crucial for extracting advantages.

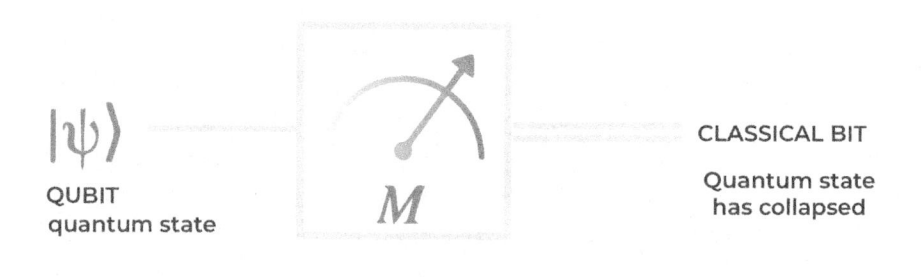

$|\psi\rangle$
QUBIT
quantum state

M

CLASSICAL BIT
Quantum state has collapsed

EMERGENCE OF CLASSICALITY

Why does our seemingly strange quantum world appear classically deterministic and predictable on everyday scales? The process of quantum decoherence causes superpositions to collapse into specific states when interacting with a noisy environment. This emergence of classicality occurs because macroscopic objects inevitably interact and become entangled with a vast number of microscopic environmental degrees of freedom. Their superpositions decohere rapidly, producing effective classical probabilistic mixtures. Quantum effects persist only for perfectly isolated microscopic particles or precisely controlled larger-scale quantum systems. Observing classical physics from the underlying quantum substrate is referred to as the quantum-to-classical transition problem. Environment-induced superselection or einselection is one proposed explanation. Interactions with the environment distinguish or "superselect" certain observable states into which superpositions decohere.

The quantum Zeno effect is another mechanism whereby frequent environmental interactions inhibit transitions between states, freezing systems in specific classical states through a type of measurement. Understanding emergent classicality remains an active area of foundational quantum physics with implications for quantum technologies.

LEGGETT-GARG INEQUALITIES

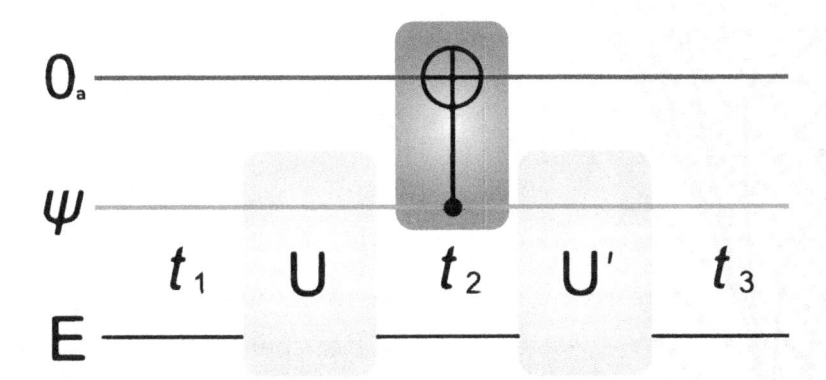

0_a ——————⊕——————

ψ ——— t_1 — U — t_2 — U' — t_3

E ——————————————

The Leggett-Garg inequalities are a set of theoretical propositions developed to test the principles of macroscopic realism and noninvasive measurability in quantum mechanics.

These inequalities were first proposed by Anthony Leggett and Anupam Garg in 1985. They aimed to explore whether the principles that apply to macroscopic objects (those we see in our everyday life) could be extended to microscopic systems (like particles in quantum mechanics).

Key Concepts of Leggett-Garg Inequalities

1. Macroscopic Realism (MR):
- This principle posits that a system with two or more macroscopically distinct states available to it will at all times be in one of those states. Essentially, it implies that physical properties exist independently of measurement.

2. Noninvasive Measurability (NIM):
   - This principle asserts that it is possible, in principle, to determine the state of a system without any disturbance to its subsequent dynamics. In the context of quantum mechanics, this means you could theoretically measure a property of a system without affecting its state.

How Leggett-Garg Inequalities Work

The Leggett-Garg inequalities involve correlations between measurements of a system's property at different times. If macroscopic realism and noninvasive measurability hold, then these correlations are constrained in certain ways which can be mathematically expressed. These constraints form the basis of the inequalities.

Mathematical Formulation

The simplest form of a Leggett-Garg inequality involves measurements at three different times, $t_1 < t_2 < t_3$. The inequality can be expressed as:

$$t_3 = t_{12} + t_{23} - t_{13} \leq 1$$

Where:
- c_{ij} is the correlation between measurements at times t_j and t_j.

If the inequality is violated, it indicates that at least one of the principles (MR or NIM) does not hold. Quantum systems often exhibit violations of these inequalities due to the inherent properties of quantum states such as superposition and entanglement.

Many experimental tests of the Leggett-Garg inequalities have been conducted, particularly with small quantum systems like photons, nuclear spins, and superconducting qubits. These experiments generally show violations of the inequalities, supporting the non-classical nature of quantum mechanics.

THE HANBURY BROWN-TWISS EFFECT

The Hanbury Brown-Twiss effect is a quantum optical phenomenon that arises due to the statistical correlations between photons in a beam of light. It can be described mathematically using the concept of second-order coherence, which is a measure of the correlation between intensities at two different points in space or time.

The second-order coherence function, denoted as $g^{(2)}(\tau)$, is defined as:

$$g^{(2)}(\tau) = \langle I(t)I(t+\tau)\rangle / \langle I(t)\rangle^2$$

where $I(t)$ is the intensity of the light at time t, τ is the time delay between the two intensity measurements, and $\langle ... \rangle$ represents the time average.

For a chaotic or thermal light source, such as a conventional light bulb or a star, the second-order coherence function takes the following form:

$$g^{(2)}(\tau) = 1 + |g^{(1)}(\tau)|^2$$

where $g^{(1)}(\tau)$ is the first-order coherence function, which describes the degree of coherence of the light field.

For a chaotic light source, the first-order coherence function $g^{(1)}(\tau)$ decays rapidly with increasing time delay τ, and the second-order coherence function approaches a constant value of 2 for large values of τ:

$$g^{(2)}(\tau \gg 0) = 2$$

This value of 2 for the second-order coherence function at large time delays is a signature of the Hanbury Brown-Twiss effect and indicates the bunching behavior of photons in a chaotic light beam.

The experimental setup used by Hanbury Brown and Twiss to observe this effect involved splitting a beam of light from a star (Sirius) into two paths and directing them onto two separate photomultiplier detectors. They measured the coincidence rate of photon detections between the two detectors as a function of the time delay between the detections.

The coincidence rate $R(\tau)$ is proportional to the second-order coherence function:

$$R(\tau) \propto g^{(2)}(\tau)$$

For a chaotic light source, they observed that the coincidence rate at zero time delay ($\tau = 0$) was higher than the coincidence rate at large time delays, indicating the bunching of photons in the light beam.

The Hanbury Brown-Twiss effect can be understood as a consequence of the quantization of the electromagnetic field and the particle nature of light. The bunching of photons arises from the fact that the detection of a photon at one point in space or time increases the probability of detecting another photon in the same vicinity, due to the indivisible nature of photons.

This effect has profound implications in various fields, including quantum optics, stellar interferometry, particle physics, and quantum information science, as it provides insights into the quantum nature of light and the fundamental properties of photons.

WHEELER'S DELAYED CHOICE EXPERIMENT

In Wheeler's thought experiment, the decision to measure a photon's path or wave-like interference is made after the photon enters the interferometer. Thus, the choice seems to retroactively determine the photon's history. These highlight peculiar effects arising from quantum measurement invasively projecting properties onto systems.

Experiments have confirmed these "delayed choice" quantum eraser effects. They force us to accept that the act of measurement irrevocably disturbs what is measured at the quantum scale. There is no objective underlying physical reality that exists independent of the measurement. Einstein viewed this invasion of physics by the observer as a defect of quantum theory. Wheeler instead saw it as a clue to the participatory nature of reality.

THE QUANTUM ZENO EFFECT

The quantum Zeno effect is a phenomenon in quantum mechanics that describes how frequent measurements or observations of a quantum system can inhibit or prevent its evolution or transition from one state to another. It is named after the ancient Greek philosopher Zeno of Elea, who proposed various paradoxes related to motion and change.

The quantum Zeno effect can be summarized as follows: If a quantum system is subjected to frequent measurements or observations, the probability of finding it in its initial state remains high, even if the system would normally evolve into a different state over time. In other words, frequent measurements can "freeze" or "lock" the system in its initial state, preventing it from undergoing the expected transitions.

Mathematically, the quantum Zeno effect can be described using the following formula:

$$P(t) = 1 - (t^2 / \tau^2) + O(t^3)$$

where P(t) is the probability of finding the system in its initial state after a time t, and τ is the characteristic time scale for the system's evolution. The higher-order terms $O(t^3)$ become negligible for small values of t.

This formula shows that as the time between measurements (t) becomes smaller and smaller, the probability P(t) of finding the system in its initial state approaches 1, effectively "freezing" the system in its initial state.

The quantum Zeno effect has practical implications in various fields, including quantum computing, where it can be used to suppress decoherence and maintain the quantum state of a system for longer periods. It also finds applications in quantum control and quantum metrology, where frequent measurements can be used to manipulate or stabilize quantum systems.

QUANTUM ERASURE

In quantum eraser experiments, entanglement is used to erase path information and recover interference between quantum states. This highlights the role of irreversible measurement collapse in destroying interference. By carefully erasing which path information using quantum correlations, superposition between states can be restored.

Quantum erasers demonstrate that loss of interference visibility is not due to disturbances in the interferometer but specifically due to measurement. Erasing information about the path retroactively appears to restore the photon's wave function to display interference again. Experiments have explored the delayed erasure of past measurements and entanglement between multiple particles exhibiting complementary behavior. Quantum erasure vividly illustrates the role of information gain versus preservation of coherence enabled by quantum correlations.

HIDDEN VARIABLE THEORIES

Dissatisfied with the indeterministic nature of quantum theory, Einstein and others hypothesized that quantum systems possess hidden variables that determine measurement outcomes. If these variables could be discovered, they hoped to formulate a deterministic theory underlying quantum mechanics. However, Bell's theorem rules out such local hidden variable interpretations.

Some non-local hidden variable theories remain viable, such as Bohmian mechanics, which adds particle trajectories to the Schrödinger equation. But these require rejecting locality or allowing faster-than-light influences. Experiments overwhelmingly support standard quantum theory over alternatives with hidden variables. Most physicists thus reject hidden variable approaches despite their appeal of restoring determinism. The probabilistic nature of quantum theory must be accepted, even if the source of that randomness remains deeply mysterious.

QUANTUM NON-DEMOLITION MEASUREMENTS

Most quantum measurements alter the state of the system observed. However, quantum non-demolition (QND) measurements minimize the backaction of measurement to avoid disrupting certain properties of interest. This requires careful coupling to only commute with observables being measured while avoiding disturbance of others.

QND schemes are used to enable continuous monitoring of fragile quantum states. This allows real-time tracking of quantum evolution and feedback control. QND measurements enable indirect observation of phenomena like quantum jumps between states or passage through avoided-level crossings. Applications include gravity wave detectors and quantum optics. QND measurements provide insights into quantum measurement theory, and help characterize quantum systems with minimal invasiveness.

Mathematical Foundations of Quantum Computing: Dirac Notation and Quantum Gates

DIRAC NOTATION FOR QUANTUM STATES

In quantum mechanics, the Dirac notation, introduced by Paul Dirac, is a compact and widely used way of representing quantum states and operators. It provides a concise and elegant way to describe the mathematical formalism of quantum theory.

The Dirac notation represents quantum states as ket vectors, denoted by $|\psi\rangle$, where ψ is a symbol representing the specific state. The corresponding bra vector, denoted by $\langle\psi|$, is the conjugate transpose (or Hermitian adjoint) of the ket vector.

The inner product between two ket vectors $|\psi\rangle$ and $|\varphi\rangle$ is represented as $\langle\psi|\varphi\rangle$, which is a complex number known as the probability amplitude. The squared modulus of the inner product, $|\langle\psi|\varphi\rangle|^2$, gives the probability of finding the system in the state $|\varphi\rangle$ when measured in the basis defined by $|\psi\rangle$.

Operators in quantum mechanics are represented using the same notation. For example, the operator $\hat{A}$ acting on the state $|\square\rangle$ is written as $\hat{A}|\psi\rangle$, which results in a new state vector.

The Dirac notation also includes useful rules and properties, such as:

1. Completeness relation: $\sum_i |i\rangle\langle i| = 1$, where $\{|i\rangle\}$ is a complete set of orthonormal basis vectors.
2. Orthogonality: $\langle\psi|\varphi\rangle = 0$ if $|\psi\rangle$ and $|\varphi\rangle$ are orthogonal states.
3. Normalization: $\langle\psi|\psi\rangle = 1$ for a normalized state $|\psi\rangle$.
4. Linearity: $\hat{A}(\alpha|\psi\rangle + \beta|\varphi\rangle) = \alpha(\hat{A}|\psi\rangle) + \beta(\hat{A}|\varphi\rangle)$, where α and β are complex numbers.

The Dirac notation simplifies the mathematical representation of quantum states and operations, making it easier to manipulate and calculate quantities of interest in quantum mechanics.

THE BLOCH SPHERE

The Bloch sphere is a geometrical representation of the pure state space of a two-level quantum system, such as a spin-1/2 particle or a qubit in quantum computing. It provides a convenient way to visualize and manipulate quantum states in a three-dimensional space. The Bloch sphere is a unit sphere, where each point on the surface or interior represents a unique quantum state of the two-level system. The north and south poles of the sphere represent the two computational basis states, often denoted as |0⟩ and |1⟩, respectively.

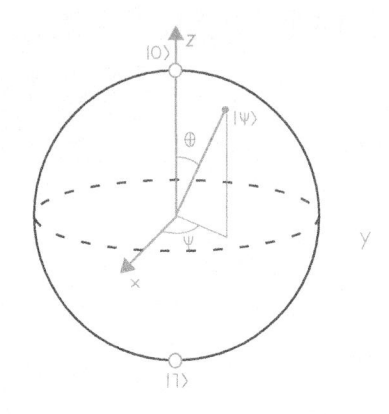

The state of the quantum system is represented by a vector, known as the Bloch vector, originating from the center of the sphere and terminating at a point on the surface or interior of the sphere. The length of the Bloch vector represents the purity of the state, with a length of 1 corresponding to a pure state and a length less than 1 corresponding to a mixed state.

The coordinates of the Bloch vector are commonly denoted as (x, y, z), where x, y, and z are real numbers satisfying the condition $x^2 + y^2 + z^2 \leq 1$. These coordinates are related to the complex coefficients (α, β) of the quantum state |ψ⟩ = α|0⟩ + β|1⟩ through the following equations:

$x = 2Re(\alpha\beta)$

$y = 2Im(\alpha\beta)$

$z = |\alpha|^2 - |\beta|^2$

The Bloch sphere provides a powerful visualization tool for understanding and manipulating quantum states. Various operations on quantum states, such as rotations, can be represented geometrically on the Bloch sphere. For example, a rotation around the z-axis corresponds to a phase shift, while rotations around the x and y axes correspond to specific quantum gates.

The Bloch sphere is widely used in quantum information theory, quantum computing, and the study of quantum systems. It helps in understanding the behavior of qubits, designing quantum algorithms, and analyzing the effects of noise and decoherence on quantum systems.

QUANTUM GATES AND CIRCUITS

How are logical operations performed on qubits? Quantum circuits consist of quantum gates that transform qubit states. Gates like Pauli-X, Hadamard, Phase, and CNOT can be represented as matrix operators acting on qubits analogous to logic gates in classical circuits. The Pauli-X gate flips or NOTs qubits, while the Hadamard gate puts qubits into balanced superpositions. Phase gates induce a relative phase between the |0⟩ and |1⟩ states. Controlled gates like CNOT flip the target qubit conditional on the control being 1. Just like classical logic gates, networks of these quantum gates are sufficient for universal quantum computation. Reversible logic helps avoid qubit measurement and decoherence during computation.

Quantum algorithms are compiled into a sequence of fundamental quantum logic gates applied to the qubits prepared in the input state. Single qubit gates like Hadamard, Phase, and Pauli gates manipulate individual qubit states. Multi-qubit gates like CNOT entangle qubit states. A universal gate set like {Hadamard, Phase, CNOT} is sufficient to approximate any unitary operation on multiple qubits to arbitrary accuracy. Circuits are designed to minimize gate count and depth to reduce errors. Reversible classical logic operations help embed computations within quantum circuits while avoiding measurement. The output state can then be measured to read out the results of the quantum computation implemented via the gate sequence.

THE CHSH GAME AND BELL'S INEQUALITY

The CHSH (Clauser-Horne-Shimony-Holt) game is a thought experiment that demonstrates the violation of Bell's inequality, which is a fundamental result in quantum mechanics that rules out certain classes of classical theories and highlights the non-local nature of quantum mechanics.

The CHSH game involves two players, Alice and Bob, who are separated and cannot communicate with each other. Each player is given a binary input (0 or 1) and must return a binary output (0 or 1) based on their input and a predetermined strategy. The goal is for

Alice and Bob to win the game by satisfying certain conditions on their outputs, which depend on both their inputs.

The rules of the CHSH game are as follows:

1. Alice receives a random input bit A (0 or 1), and Bob receives a random input bit B (0 or 1).

2. Alice and Bob must independently output bits a and b, respectively, based on their inputs and a pre-agreed strategy.

3. The players win the game if a ⊕ b = A · B, where ⊕ denotes addition modulo 2, and · denotes multiplication modulo 2.

The remarkable result is that if Alice and Bob use a classical strategy (i.e., a local hidden variable theory), their maximum probability of winning the CHSH game is 0.75 or 75%. This limit is known as the Bell inequality or the CHSH inequality.

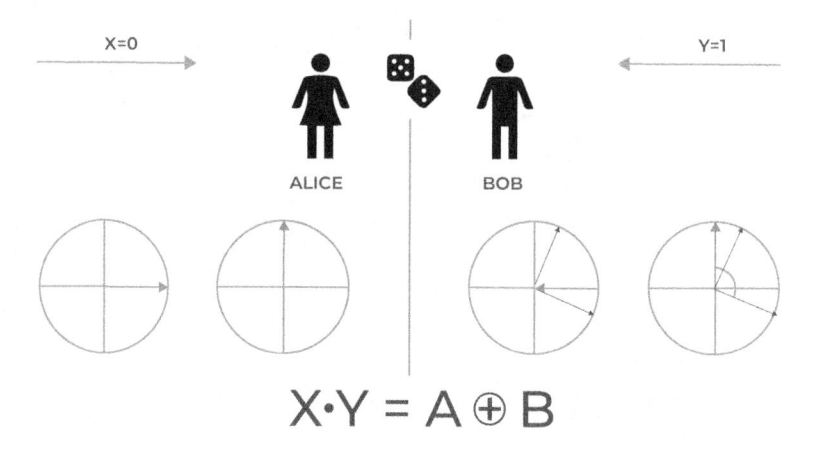

$$X \cdot Y = A \oplus B$$

However, if Alice and Bob employ a quantum strategy, they can achieve a winning probability of up to $\cos^2(\pi/8) \approx 0.85$ or 85%, violating the Bell inequality. This violation is achieved by using a specific quantum state (e.g., the Bell state) and performing appropriate measurements on their respective qubits. The violation of Bell's inequality has profound implications for our understanding of quantum mechanics and the nature of reality. It rules out the possibility of explaining quantum phenomena using local hidden variable theories, which assume that particles carry pre-determined values for all their properties and that measurements merely reveal these values. Instead, the violation of Bell's inequality supports the non-local nature of quantum mechanics, where entangled particles exhibit correlations that cannot be explained by any local realistic theory. This non-locality challenges our classical intuitions and has led to intense debates and discussions about the fundamental principles of quantum mechanics and the nature of reality.

QUANTIZING CLASSICAL SYSTEMS

How do we model real-world systems quantum mechanically? Physical observables like position, momentum, and energy become quantum operators with discrete spectra. Phenomena like spin, orbital angular momentum, and polarization arise from quantum descriptions of matter. Systems, once treated classically, can display new quantum effects when quantized.

Quantum simulations apply these quantization prescriptions to systems such as chemical interactions, fluid dynamics, electromagnetism, and gravitation. Digital quantum computers approximate the resulting quantum models. Analog quantum simulators more directly embody them using tailored hardware like ultracold atoms, trapped ions, or photons. Comparing simulated predictions with experiments provides insights into quantum mechanics, computational complexity, and properties of exotic quantum materials.

COMPATIBLE OBSERVABLES AND DEGENERATE CODESPACES

Compatible observables are quantum observables (operators representing physical quantities) that can be measured simultaneously with arbitrary precision. In other words, if two observables are compatible, it is possible to find a common set of eigenvectors that are simultaneous eigenvectors of both observables.

Mathematically, two observables represented by Hermitian operators $\hat{A}$ and $\hat{B}$ are compatible if and only if they commute:

$[\hat{A}, \hat{B}] = \hat{A}\hat{B} - \hat{B}\hat{A} = 0$

If two observables commute, their corresponding operators share a common set of eigenvectors, and their measurements can be performed simultaneously without disturbing each other's outcomes.

Degenerate codespaces are subspaces of the Hilbert space (the vector space representing the state space of a quantum system) that are invariant under the action of a set of commuting observables. These subspaces are spanned by the simultaneous eigenvectors of the commuting observables, and they represent the possible states of the system that are compatible with the given set of observables.

The concept of degenerate codespaces is particularly relevant in the context of quantum error correction and fault-tolerant quantum computation. In these contexts, degenerate codespaces are used to encode quantum information in a way that is resistant to certain types of errors or noise.

Consider a set of commuting observables $\{\hat{O}\_1, \hat{O}\_2, ..., \hat{O}\_n\}$, which represent the errors or noise that can affect the quantum system. The degenerate codespaces are the subspaces of the Hilbert space that are invariant under the action of these observables, meaning that any state within a degenerate codespace remains within that codespace under the action of the error operators.

By encoding quantum information in these degenerate codespaces, it is possible to protect the information against the errors or noise represented by the commuting observables. This is because any error operator that commutes with the encoded state leaves the state unchanged (up to a global phase factor).

The mathematical description of degenerate codespaces involves finding the simultaneous eigenvectors and eigenspaces of the commuting observables. These eigenvectors and eigenspaces form the basis for the degenerate codespaces, and they can be used to construct error-correcting codes or fault-tolerant quantum circuits.

Degenerate codespaces play a crucial role in the development of practical quantum computing systems, as they provide a way to mitigate the effects of noise and errors, which are major challenges in the realization of large-scale quantum computers.

QUANTUM DISCORD AND INFORMATION

Quantum discord measures quantum correlations beyond just entanglement, including those arising in certain mixed separable states. Unlike classical correlations, discord depends on the order of measurement. Discord captures the loss of coherence caused by local measurements. This reveals quantum advantages that are possible even without full entanglement.

Discord indicates the presence of quantum correlations useful for quantum computing protocols. Almost all quantum states have some discord allowing extraction of quantum advantages. Studies suggest even low-discord states can display speedups over classical. Discord is also connected to cryptography and state merging. Ongoing research aims to develop discord-based benchmarks of quantumness complementary to entanglement measures. Operationally utilizing discordant states could enable quantum information processing in regimes that are not possible with entanglement alone.

QUANTUM HARDWARE FOR COMPUTATION AND SIMULATION

Several physical systems are promising candidates for acting as qubits and platforms for quantum information processing. These include superconducting circuits, trapped ions, ultracold neutral atoms, photonics, quantum dots, defect centers, and topological systems. Each has strengths and challenges regarding coherence, control, coupling, and scalability.

Ongoing research aims to practically realize stable, low-noise qubits with efficient interconnects while minimizing manufacturing complexity. Hybrid approaches combining multiple physical qubits and platforms could prove advantageous. Tailoring hardware to specific applications like quantum chemistry, optimization, or machine learning is also being explored. Advancing qubit coherence, gate fidelities, and manufacturing scalability remain key near-term goals on the path to full-fledged quantum computers and simulators.

PHOTONIC QUANTUM COMPUTING

Photons are appealing for their low decoherence, ease of transmission over long distances, and ability to leverage existing optical infrastructure. Linear optics provides a starting point, but full quantum computing requires nonlinear interactions between photons. This can be achieved using measurement-induced nonlinearities or materials with strong photon coupling.

Integrated photonics enable miniaturized circuits to manipulate the quantum states of light. Multiple photons can also be generated on a chip, enabling photon-photon entangling gates. Further improving these entangling interactions and incorporating quantum error correction remains necessary for optical quantum computing.

Hybrid approaches using photons coupled to matter qubits are also promising. Photonic qubits are strong candidates for quantum communication and distributed quantum networks.

TOPOLOGICAL QUANTUM COMPUTING

Topological qubits are encoded in nonlocal properties of materials rather than microscopic states at specific locations. This provides inherent protection against local perturbations that cause decoherence in conventional qubits. Topological quantum computing aims to harness these robust, stable correlations for inherently fault-tolerant information processing.

Quasiparticles known as non-Abelian anyons have exotic statistics suitable for topological qubits. Majorana zero modes in certain superconductors are predicted to exhibit these properties. Braiding anyons could implement logic gates in a manner resilient to external noise.

Although challenges remain in unambiguously detecting and manipulating anyons, proof-of-principle demonstrations have renewed interest in this approach. Topological hardware would synergistically combine with error correction software for maximal robustness.

ADIABATIC QUANTUM COMPUTING

Instead of circuit-based gate operations, adiabatic quantum computing relies on adiabatically evolving a Hamiltonian to keep qubits in the ground state. The final Hamiltonian is designed such that its ground state encodes the solution to a problem. This quantum annealing approach does not require detailed control of individual gates and qubits.

The adiabatic theorem guarantees a system initialized in the ground state will remain there if the Hamiltonian is varied slowly enough. In practice, finite temperature causes decoherence out of the ground state, limiting this process. Quantum tunneling also enables shortcutting between local minima. Quantum annealers built on superconducting qubits currently have limited coherence, restricting problem complexity. Improving thermal isolation may enable larger-scale adiabatic quantum optimization.

QUANTUM NEURAL NETWORKS AND MACHINE LEARNING

Quantum machine learning explores the use of quantum systems for pattern recognition, classification, learning, and artificial intelligence. Quantum neural networks aim to utilize quantum effects like superposition and entanglement to improve neural models for machine learning tasks. Hybrid classical quantum machines combine quantum processors with classical learning algorithms.

An active area of research focuses on quantizing classical neural network components like weights, activations, and objective functions. Training the resulting quantum neural networks remains challenging. Variational quantum circuits using parameterized gates are a promising approach, with demonstrations already showing quantum advantages for some learning problems. Looking ahead, deeper integrations of quantum physics with learning theory could profoundly advance artificial intelligence.

DIGITAL VS ANALOG QUANTUM SIMULATION

Quantum simulations use controllable quantum systems to model other less accessible quantum systems for insights into their properties. Digital quantum simulation approximates the evolution via a sequence of discrete gate operations. Analog simulation directly mimics the target evolution Hamiltonian in a dedicated hardware platform like ultracold atoms.

Digital quantum computers have flexible programmability, allowing for the simulation of many different physical systems. But gate errors accumulate over long evolutions, limiting performance without error correction. Analog simulators tailored to specific models can precisely match complex quantum interactions and dynamics. This avoids gate errors but lacks flexibility across simulation problems. Hybrid approaches combining analog quantum hardware with programmable digital control aim to get the best of both approaches.

DECOHERENCE-FREE SUBSPACES AND NOISELESS SUBSYSTEMS

Special quantum error correction encodings allow the construction of decoherence-free subspaces and noiseless subsystems. Qubits encoded this way are completely decoupled from certain types of environmental noise by careful engineering of symmetries. This passive approach to avoid errors complements active quantum error correction. Decoherence-free spaces employ compatible observables and selective couplings to "hide" quantum information orthogonal to the noise. Noiseless subsystems use block-diagonal algebra to partition the system into sectors unaffected by noise. Although these techniques are limited, they illustrate the rich possibilities of encoding information in carefully designed quantum state space structures.

QUANTUM MONEY AND CRYPTOCURRENCIES

Quantum money uses quantum effects like entanglement to create cash-like tokens that are physically impossible to counterfeit or copy. This has applications in cryptography, digital payments, and cryptocurrencies secured by the laws of physics.

Proposed quantum money schemes rely on fundamental results like the no-cloning theorem to prevent duplication. Verification of quantum banknotes can be done via cryptographic protocols. Quantum money that can be forged but not copied efficiently also has uses as a cryptographic primitive. Real-world quantum money systems will require overcoming technological challenges in manufacturing, storing, and transmitting complex entangled quantum states at scale.

Quantum Computing as a Paradigm Shift: Implications and Challenges

EXPONENTIAL SPEEDUPS FROM QUANTUM PARALLELISM

What makes quantum computers so powerful? The ability to represent information in quantum superpositions allows for massive parallelism during computation since a system of N qubits exists in all 2N states simultaneously. Consider a search problem where we classically try N solutions one by one, taking N steps. A quantum algorithm like Grover's algorithm could achieve the same using just sqrt(N) steps, providing a quadratic advantage. Factoring large numbers using Shor's algorithm gives an exponential speedup over classical factoring. As we scale the number of qubits, the parallelism grows exponentially - enabling astounding speeds for certain problems.

Quantum parallelism arises from the ability of qubits to exist in superpositions of 0 and 1. While a classical n-bit register can represent one of 2n possible numbers at a time, a quantum system of n qubits represents all 2n numbers in superposition. Interference between these parallel states enables certain problems like factoring, search, simulation, and optimization to be solved with substantially fewer steps than classically needed. A quantum computer with just 50 qubits could consider over a quadrillion values simultaneously in parallel, exceeding the memory capacity of any existing supercomputer. However, we cannot directly observe all these parallel states. Clever quantum algorithms

LIMITS DUE TO NON-DETERMINISTIC OUTPUTS:

However, there are some subtleties to quantum parallelism. Unlike a classical parallel processor, we cannot directly read all the outputs in a superposition. Upon measurement, the quantum state collapses randomly to a single result. So, quantum algorithms must be carefully constructed to still take advantage of the inherent parallelism. Techniques like amplitude amplification, phase estimation, and quantum Fourier transforms help extract useful information efficiently from the parallel computations occurring in superposition. Still, quantum computers will not provide speedups for all tasks and face several challenges.

Directly measuring a superposition state provides a random bit string, not the desired result. The non-deterministic readout of the parallel outputs appears to negate the quantum advantage. However, by using interference to modify the amplitudes before measurement, we can boost the probability of measuring a "good" state that provides the solution. But algorithms must be designed such that this works with a feasibly small number of runs. If exponentially many measurements are needed, the quantum speedup disappears. So, while quantum parallelism enables tremendous computational power, skillfully extracting meaningful results remains challenging in practice.

KEY CHALLENGES: DECOHERENCE, NOISE, AND ERROR CORRECTION:

A key challenge is that quantum superpositions are extremely fragile. Any slight interaction with the environment collapses the superposition to a single state, an effect called quantum decoherence or dephasing. Qubits also suffer errors during logic gate operations due to environmental noise. The fragility of quantum information requires quantum error-correcting codes that use many physical qubits to redundantly encode each logical qubit. Error correction imposes steep physical qubit overheads. For example, estimates suggest that reliably implementing a 32-logical qubit system may require millions of noisy physical qubits - a daunting challenge. Much work remains both in hardware design and quantum software to overcome these hurdles.

Quantum decoherence quickly causes superposition states to decay into incoherent statistical mixtures. Gate operations on qubits are also noisy, which introduces errors during algorithm execution. Together, these effects accumulate, limiting the size of quantum circuits that can be run reliably. Quantum error correction provides a solution - but requires logical qubits to be encoded in a spread-out, redundant fashion across many physical qubits using quantum coding techniques. The massive overhead ratios involved are sobering - but theoretically show a path to fault tolerance. Continued progress in cutting-edge hardware, error-mitigating software, and encoding efficiency is critical to overcoming the fundamental challenge of fighting environmental noise and decoherence.

THE THREAT TO ENCRYPTION AND CRYPTOCURRENCIES:

The ability of quantum computers to crack encryption schemes like RSA and elliptic curve cryptography quickly poses security threats to our communications infrastructure and financial systems. Many modern cryptographic protocols rely on the computational difficulty of finding the prime factors of large numbers or computing discrete logarithms. Shor's algorithm can break both these problems in polynomial time, breaking widely used public key encryption and digital signature schemes. Cryptocurrency systems also rely on similar cryptography and face threats from future quantum capabilities. Upgrading to quantum-resistant encryption and cryptocurrencies will be critical.

Quantum computers threaten the security assumptions underlying the public key infrastructure powering secure Internet and financial transactions worldwide. Public keys are intended to be impossible for others to figure out - but can be easily deduced by a quantum computer running Shor's algorithm. This jeopardizes information protected using RSA and related encryption schemes. Cryptocurrencies like Bitcoin also rely on cryptographic hashing, which could potentially be broken by quantum cryptanalysis. Transitioning classical cryptography to new quantum-robust schemes like lattice-based cryptography is necessary to safeguard modern computing systems in the emerging quantum era.

CONCERNS AROUND EQUITY, ETHICS, AND MISUSE:

The significant resources and expertise required to build quantum computers raise concerns about equitable access to quantum capabilities. Like any powerful technology, quantum systems also face risks of misuse by malicious actors and adversarial nations. There are calls for increased oversight into the ethical use and democratization of quantum computing. On the positive side, quantum techniques applied to machine learning, chemistry, and medicine hold tremendous potential for advancing human welfare. Public policy discussions around the responsible use of quantum computing will increase as the field matures.

Experts warn that quantum computing breakthroughs may widen technology disparities between rich and poor nations. The potential for quantum-enabled decryption, cyberattacks, and surveillance also raises alarms. For such a disruptive technology, reasoned debate on responsibility, ethical guidelines, and democratization is necessary. The quantum community has an obligation to ensure powerful capabilities are used equitably for human betterment, not destruction. Quantum insights into nature could also profoundly shift human worldviews in unpredictable ways. Navigating these societal impacts requires an interdisciplinary outlook spanning computer science, ethics, policy, and the humanities.

CHAPTER 3

QUANTUM HARDWARE OVERVIEW

Quantum Bits (Qubits) and Their Properties

BIT Classical Computing

QUBIT Quantum Computing

QUBIT DEFINITION

The fundamental unit of quantum information is the quantum bit or qubit. Unlike classical bits limited to binary 0 or 1 states, qubits can exist in a superposition of 0 and 1. This allows qubits to encode information in a fundamentally unique way from classical bits, enabling new computing capabilities. The quantum properties of superposition, entanglement, and interference open up powerful ways to manipulate and process information.

Qubits are the basic building blocks used to construct quantum computers; just as classical bits are the basic units of classical computers. But qubits harness quantum physics phenomena not exhibited by classical bits. Thanks to superposition, a qubit can represent a 0, a 1, or, crucially, any linear combination of both simultaneously. Entangled qubits have correlated states that are not possible with classical bits. Quantum algorithms manipulate these qubit properties to achieve computational speedups over their classical counterparts for certain problems. The engineering of high-quality qubits is crucial to realizing scalable, reliable quantum computation.

REPRESENTING QUBITS ON THE BLOCH SPHERE

The state of a single qubit finds visualization on the surface of a unit 3D sphere known as the Bloch sphere. The north and south poles of this sphere represent the basis states 0 and 1. All other points on the Bloch sphere correspond to superpositions of 0 and 1, described by two angles. This geometric representation offers an intuitive means to comprehend single qubit states and their evolution under operations.

The Bloch sphere serves as a convenient geometrical representation of a qubit's state. Pure states align with points on the sphere's surface, while mixed states occupy the interior volume. Executing a single qubit gate results in the rotation of the state vector to a new point on the sphere. The use of two angles to parameterize rotations provides a straightforward visualization of qubit manipulations. The distance between points indicates the overlap between quantum states. While the Bloch sphere proves useful for reasoning about single qubit dynamics, visualizing entangled multi-qubit states necessitates transitioning to a higher-dimensional Hilbert space representation beyond the 3D Bloch sphere.

QUBIT PROPERTIES

Qubits derive their computational power from fundamental quantum properties—superposition, entanglement, and interference—absent in classical analogs. Superposition empowers a qubit to exist in all possible states simultaneously. Entanglement establishes correlations between states of separated qubits in a non-classical manner. Interference generates probabilistic amplitudes that can be either constructive or destructive. Together, these non-intuitive quantum effects enable potent quantum computation.

While a classical bit must be either 0 or 1, a qubit can exist in a superposition like $α|0> + β|1>$, where $α$ and $β$ are probability amplitudes. This ability to encode 0 and 1 states simultaneously underpins the massive parallelism of quantum computers for specific problems. Entanglement further distinguishes quantum systems; when two qubits are entangled, their states become interdependent, irrespective of physical separation. Interference enables the reinforcement or cancellation of paths, concentrating the probability of measuring a desired state. Superposition, entanglement, and interference are pivotal to most quantum algorithm speedups. Quantum hardware must maintain these fragile quantum states long enough to perform useful computations.

QUBIT FRAGILITY AND DECOHERENCE

A major hardware challenge is that qubits are fragile - interaction with the environment collapses superposition states into classical mixtures through quantum decoherence. Precision quantum engineering is needed to minimize noise and maintain coherence long enough to realize quantum advantage. This motivates operating at extremely cold temperatures and exploring hardware options like topological qubits more intrinsically robust to errors.

Unlike classical bits, the quantum nature of qubits makes them highly sensitive to perturbations from the surrounding environment. Any slight interaction can cause decoherence that collapses superposition states into classical probabilistic mixtures. This quickly destroys the quantum advantage. Carefully engineered qubits stabilized through techniques like cryogenic operation, materials selection, pulse shaping, and error correction codes are required to reduce decoherence. But some residual decoherence is inevitable, imposing limits on algorithm runtimes, qubit connectivity, gate error rates, and quantum memory lifetimes. Overcoming decoherence remains one of the greatest hardware challenges for scalable quantum computing.

READOUT AND MEASUREMENT OF QUBITS

Reading out qubit states at the end of a quantum algorithm is challenging because measurement inherently disturbs quantum systems. Techniques like amplifying signals, using ancilla qubits, and rapidly repeating readouts to gain statistics help infer qubit states from noisy measurement data. High-fidelity single-shot readout of individual qubits with minimal disturbance remains an active research target.

Projective measurement of a qubit on a 0/1 basis will randomly collapse the state while providing just a single bit of classical information. Gaining more insights into qubit states requires strategies like weak measurements, partial collapse, using ancilla qubits, and quantum non-demolition approaches that minimize backaction. Repeating rapid state collapse and readout thousands of times yields statistics revealing the probability distribution of qubit states prior to measurement. Improving the speed, fidelity, and scalability of quantum state readout is essential for extracting useful results from quantum algorithms and error correction protocols.

QUBIT CONNECTIVITY AND CONTROL

Practical quantum computation necessitates the controllable interaction of multiple qubits according to programmed sequences. Connecting qubits to interact and correlate their states is essential for implementing quantum logic gates. However, adding connections also expands avenues for decoherence and noise. Optimizing qubit connectivity graphs for specific algorithms involves balancing tradeoffs between gate complexity and error rates.

Physically coupling qubits enables multi-qubit gates like CNOT, which entangle states to achieve speedups. However, connecting qubits also exposes them to cross-talk errors, degrading performance. Limiting connectivity reduces gate complexity but constrains algorithm capabilities. Hardware engineers continue to improve interconnect technologies like flip-chip integration of multiple superconducting qubits on the chip. Nevertheless, intrinsic limits on how many neighboring interactions can be supported before noise accumulates remain a challenge. Novel architectures with modular local units coupled through photonic channels provide an alternative path.

QUBIT FIDELITY METRICS

Quantifying qubit performance metrics such as state preparation fidelity, gate fidelity, coherence times, and readout fidelity offers standard benchmarks for comparing hardware implementations. As technology improves, these fidelity numbers help track progress. However, translating fidelities into practical algorithm performance also requires a holistic analysis of error accumulation and correction. Qubit fidelities are steadily improving across technologies like superconducting circuits, trapped ions, and quantum dots. Key metrics include gate error rates approaching the 10^{-9} to 10^{-12} range needed for fault tolerance, coherence times nearing milliseconds, and readout fidelities exceeding 99%. However, these individual qubit metrics do not directly reveal the algorithmic power of a system. Additional factors like interconnect flexibility, classical control integration, and error correction overheads determine practical computation capabilities. As quantum technologies mature, developing realistic performance benchmarks will require co-design spanning across low-level physics up through applications.

COST AND SCALABILITY OF QUBIT TECHNOLOGIES

Practical applications demand quantum computers with large numbers of qubits. However, scaling up qubit counts while maintaining control precision and coherence is extremely challenging. Relative scalability is an important consideration when comparing qubit modalities. Lower cost per qubit also accelerates development and access. Estimating realistic qubit manufacturing costs at scale guides technology roadmaps and commercial viability.

Most existing quantum computers have 10-100 qubits. Realizing advantages for applications like quantum chemistry or machine learning is anticipated to require upwards of thousands of logical qubits. This imposes steep demands on improving physical qubit performance, error correction overhead, and connectivity. Some qubit technologies like ion traps or quantum dots have higher complexity per qubit but support shorter-range interactions. The cost of classical control electronics can also scale poorly. Balancing precision, qubit count, and manufacturability constraints poses significant technical and commercial challenges on the path to pragmatic large-scale quantum computing.

Quantum Hardware Architectures: Superconducting Qubits, Ion Traps, and Topological Qubits

SUPERCONDUCTING QUBITS

Superconducting circuits cooled to near absolute zero temperature exhibit quantum behavior exploitable as qubits. Josephson junctions, loops, and resonators can be microfabricated using standard lithography to create integrated circuits acting as qubits with controllable interactions. This leverages existing semiconductor manufacturing infrastructure.

Macroscopic superconducting wires have properties that allow their quantum state to be manipulated as qubits. Josephson junctions act as nonlinear circuit elements, enabling everything from charge, flux, and phase qubits to exotic variants like 0-Pi qubits. Planar integrated wiring patterns determine qubit interactions. While coherence times are shorter than other qubit options, the prospect of mass-producible quantum microchips has made superconducting circuits a leading candidate for realizing commercial quantum computing technology. Ongoing challenges include boosting qubit lifetimes and fidelity while developing fabrication recipes scalable to mass manufacture.

TRAPPED ION QUBITS

Trapped ion qubits are another leading platform for quantum computing, based on the quantum states of individual ions confined in electromagnetic traps. Trapped ion qubits have several advantages, such as long coherence times, high-fidelity quantum gates, and natural connectivity between qubits, making them a promising candidate for scalable quantum computers.

In a trapped ion quantum processor, ions (typically atomic species such as Ca+, Sr+, or Yb+) are confined in a linear Paul trap or a Penning trap, using a combination of static and oscillating electric fields. The quantum states of the ions, such as the ground state $|0\rangle$ and the excited state $|1\rangle$ of a selected atomic transition (e.g., the S-D transition in Ca+), serve as the basis states of the qubits.

The ions are cooled to their motional ground state using techniques such as Doppler cooling and sideband cooling, which reduce their vibrational motion to the quantum regime. The ions can then be manipulated using laser pulses or microwave fields, which drive transitions between the qubit states and the motional states of the ions.

The Hamiltonian of a trapped ion qubit can be written as:

$$H = \omega_0 \sigma_z/2 + \sum_k \omega_k a_k^\dagger a_k + \sum_k g_k (a_k^\dagger + a_k)(\sigma_+ + \sigma_-)$$

where ω_0 is the frequency of the qubit transition, σ_z, σ_+, and σ_- are the Pauli matrices, ω_k and a_k are the frequency and the annihilation operator of the k-th motional mode, and g_k is the coupling strength between the qubit and the k-th mode.

The first term in the Hamiltonian represents the internal energy of the qubit, the second term represents the energy of the motional modes (phonons), and the third term represents the interaction between the qubit and the motional modes, which enables the implementation of quantum gates.

Single-qubit gates, such as rotations around the X, Y, or Z axis of the Bloch sphere, can be implemented by applying laser pulses or microwave fields that are resonant with the qubit transition. For example, a rotation around the X-axis by an angle θ can be described by the unitary operator:

$$R_x(\theta) = \exp(-i\theta \sigma_x/2) = \cos(\theta/2) I - i \sin(\theta/2) \sigma_x$$

where I is the identity matrix and σ_x is the Pauli-X matrix.

Two-qubit gates, such as the controlled-NOT (CNOT) gate or the Mølmer-Sørensen (MS) gate, can be implemented by exploiting the collective motion of the ions in the trap. For example, the MS gate can be described by the unitary operator:

$$U_{MS}(\theta) = \exp(-i\theta/2\, S_x^2)$$

where $S_x = \sum_i \sigma_x^{(i)}$ is the collective spin operator along the X-axis, and the angle θ depends on the intensity and duration of the laser pulses.

TOPOLOGICAL QUBITS

Topological qubits are a type of quantum bit that relies on the topological properties of certain materials, such as topological insulators or superconductors, to protect the quantum information from decoherence and errors. Unlike conventional qubits, which encode information in the physical states of a system (e.g., the charge or flux states of a superconducting circuit), topological qubits encode information in the global properties of the system, such as the braiding of anyons or the winding of Majorana fermions. The key idea behind topological qubits is that the quantum information is stored in a non-local, degenerate ground state of the system, which is protected by a gap in the energy spectrum. This gap separates the ground state from the excited states and suppresses the transitions between them, making the topological qubits resilient to local perturbations and noise.

One of the most promising candidates for topological qubits are the Majorana fermions, which are hypothetical particles that are their own antiparticles. Majorana fermions can emerge as quasiparticle excitations in certain topological superconductors, such as the $\nu=5/2$ fractional quantum Hall state or the proximitized semiconductor nanowires.

The Hamiltonian of a one-dimensional topological superconductor can be written as:

$$H = -\mu \sum_j c_j^\dagger c_j - t \sum_j (c_j^\dagger c_{j+1} + c_{j+1}^\dagger c_j) + \Delta \sum_j (c_j c_{j+1} + c_{j+1}^\dagger c_j^\dagger)$$

where μ is the chemical potential, t is the hopping amplitude, Δ is the superconducting gap, and $c_j^\dagger$ and c_j are the creation and annihilation operators of an electron at site j.

This Hamiltonian can be rewritten in terms of the Majorana operators γ_j^A and γ_j^B, which satisfy the anticommutation relations $\{\gamma_i^\alpha, \gamma_j^\beta\} = 2\delta_{ij} \delta_{\alpha\beta}$:

$$H = i\mu \sum_j \gamma_j^A \gamma_j^B - it \sum_j (\gamma_j^B \gamma_{j+1}^A - \gamma_j^A \gamma_{j+1}^B) - \Delta \sum_j (\gamma_j^A \gamma_{j+1}^A + \gamma_j^B \gamma_{j+1}^B)$$

In this representation, the Majorana operators can be thought of as the real and imaginary parts of the electron operator: $c_j = (\gamma_j^A + i\gamma_j^B)/2$.

The topological phase of the system is characterized by the presence of unpaired Majorana fermions at the ends of the nanowire, which can be combined to form a non-local fermionic state that encodes a qubit. The quantum information is protected by the spatial separation of the Majorana fermions and the energy gap of the bulk states.

To perform quantum operations on topological qubits, one can use the braiding of the Majorana fermions, which corresponds to the exchange of their positions in real space. The braiding of Majorana fermions realizes a unitary transformation on the qubit state, which depends only on the topology of the braiding path and not on the details of the system.

For example, the braiding of two Majorana fermions y_1 and y_2 can be described by the unitary operator:

$U\_\{12\} = \exp(\pi\ y\_1\ y\_2\ /\ 4) = (1 + y\_1\ y\_2)\ /\ \sqrt{2}$

which corresponds to a rotation by $\pi/2$ around the X-axis of the Bloch sphere.

PHOTON QUBITS

Photonic qubits are a type of quantum bit that uses the quantum states of light, such as the polarization or the spatial mode of a single photon, to encode and process quantum information. Photonic qubits have several advantages over other types of qubits, such as the ability to travel long distances without decoherence, the compatibility with existing optical communication networks, and the potential for room-temperature operation.

The quantum state of a single photon can be described by a superposition of two orthogonal basis states, such as the horizontal (H) and vertical (V) polarization states:

$|\psi\rangle = \alpha\ |H\rangle + \beta\ |V\rangle$

where α and β are complex amplitudes that satisfy $|\alpha|^2 + |\beta|^2 = 1$.

The Hamiltonian of a single photon in a polarization qubit can be written as: $H = \omega\ (|H\rangle\langle H| + |V\rangle\langle V|)$

where ω is the frequency of the photon.

Single-qubit gates on photonic qubits can be implemented using linear optical elements, such as wave plates and polarizing beam splitters. For example, a rotation around the X-axis of the Bloch sphere can be realized by a half-wave plate (HWP) with its fast axis at an angle θ/2 with respect to the horizontal axis:

$R_x(\theta) = \exp(-i\theta\sigma_x/2) = \cos(\theta/2) I - i\sin(\theta/2)(|H\rangle\langle V| + |V\rangle\langle H|)$

where I is the identity matrix and σ_x is the Pauli-X matrix.

Two-qubit gates on photonic qubits, such as the controlled-NOT (CNOT) gate or the controlled-phase (CZ) gate, can be implemented using non-linear optical elements, such as the Kerr effect or the Rydberg blockade, or using linear optics with ancillary photons and measurements. One of the key challenges in photonic quantum computing is the lack of deterministic two-qubit gates, which limits the scalability of the system. This challenge can be addressed by using measurement-based quantum computing, where the quantum information is processed by a sequence of single-qubit measurements on a highly entangled cluster state of photons.

QUANTUM DOT QUBITS

Quantum dot qubits are a type of quantum bit that uses the charge or spin states of electrons or holes confined in semiconductor nanostructures, called quantum dots, to encode and process quantum information. Quantum dots are nanoscale regions of a semiconductor material, such as gallium arsenide (GaAs) or silicon (Si), that can trap and manipulate individual electrons or holes.

The quantum state of an electron in a quantum dot can be described by a superposition of two basis states, such as the spin-up ($|\uparrow\rangle$) and spin-down ($|\downarrow\rangle$) states:

$|\psi\rangle = \alpha|\uparrow\rangle + \beta|\downarrow\rangle$

where α and β are complex amplitudes that satisfy $|\alpha|^2 + |\beta|^2 = 1$.

The Hamiltonian of an electron spin qubit in a quantum dot can be written as:

$H = g\mu_B B \cdot S + \Sigma_i A_i I_i \cdot S$

where g is the electron g-factor, μ_B is the Bohr magneton, B is the external magnetic field, S is the electron spin operator, A_i is the hyperfine coupling constant, and I_i is the nuclear spin operator of the i-th nucleus in the quantum dot.

The first term in the Hamiltonian represents the Zeeman energy of the electron spin in the external magnetic field, which splits the spin-up and spin-down states. The second term represents the hyperfine interaction between the electron spin and the nuclear spins in the quantum dot, which can cause decoherence and errors in the qubit. Single-qubit gates on quantum dot qubits can be implemented using microwave or laser pulses that drive transitions between the spin states. For example, a rotation around the X-axis of the Bloch sphere can be realized by a microwave pulse with a frequency that matches the Zeeman splitting of the spin states:

$R_x(\theta) = \exp(-i\,\theta\,\sigma_x/2) = \cos(\theta/2)\,I - i\sin(\theta/2)\,(|\uparrow\rangle\langle\downarrow| + |\downarrow\rangle\langle\uparrow|)$

where I is the identity matrix and σ_x is the Pauli-X matrix.

Two-qubit gates on quantum dot qubits, such as the controlled-NOT (CNOT) gate or the controlled-phase (CZ) gate, can be implemented using the exchange interaction between the electrons in nearby quantum dots, or using the dipole-dipole interaction between the electron spins.

HYBRID QUANTUM SYSTEMS

Rather than relying on a single qubit modality, hybrid systems aim to combine multiple quantum technologies to harness their relative strengths. For instance, using different physical qubits for memory, processing, and interconnects could improve capabilities compared to any one approach alone.

Superconductors or semiconductor qubits provide compact processing qubits, while ions, atoms, or spins represent good long-lifetime quantum memory units. Photons enable low-loss transmission of quantum states between them. By integrating different qubits and platforms, hybrid schemes aim to balance factors like scalability, fidelity, speed, and connectivity. This adds engineering complexity but holds promise for realizing the full computational potential of quantum technologies. The modular nature also provides built-in extensibility to incorporate emerging new qubit types as the field progresses.

Quantum Error Correction Techniques: Quantum Codes and Error Mitigation

SOURCES OF PHYSICAL QUBIT ERRORS

All physical qubit implementations suffer from imperfections that introduce errors during initialization, logic gates, memory, and measurement. Main error sources include decoherence, crosstalk, manufacturing variations, and control electronic noise. Together, these make practical error rates orders of magnitude too high for robust quantum computing.

Quantum states are inherently fragile. Qubit errors arise from decoherence, gate imperfections, fabrication variations, qubit crosstalk, control electronics, radiation events, thermal effects, and other nonidealities. At temperatures above absolute zero, entropy causes gradual leakage out of qubit basis states. Collectively, these numerous error mechanisms result in error rates around 10^{-3} to 10^{-2} per gate - far exceeding fault tolerance thresholds around 10^{-9} to 10^{-12}. Quantum error correction and fault tolerance techniques aim to effectively reduce these physical error rates to enable reliable large-scale quantum computation.

QUANTUM ERROR CORRECTION PRINCIPLES

Quantum error correction shields logical qubit states by redundantly encoding each state across multiple physical qubits. Measurements detect bit-flip and phase-flip errors, which can then be corrected through appropriate recovery operations. This process repeats recursively across logical levels, adhering to the threshold theorem to drive error rates arbitrarily low at the algorithmic level.

Analogous to classical error correction using redundant bits, quantum error correction involves repeatedly measuring multi-qubit stabilizer generators to detect errors, followed by corrective recovery gates. The additional qubits provide redundancy to overcome the no-cloning limit. As long as the physical qubit error rate stays below a critical threshold, logical error rates can be successively reduced by concatenating quantum error correcting codes in multiple layers. In principle, this approach allows for asymptotically reaching arbitrary reliability. The challenge lies in realizing quantum error correction efficiently with minimal resource overheads on noisy real hardware.

QUANTUM ERROR CORRECTION CODE EXAMPLES

Quantum error correction is a critical component of building reliable and fault-tolerant quantum computers. Quantum error correction codes are designed to detect and correct errors that occur during quantum computations, which can be caused by various sources such as decoherence, noise, and imperfect control. In this section, we will discuss some examples of quantum error correction codes and their properties.

Bit-Flip Code (Three-Qubit Code):

The bit-flip code is one of the simplest quantum error correction codes, designed to correct single-qubit bit-flip errors (i.e., errors that change the state from |0⟩ to |1⟩ or vice versa). The code encodes one logical qubit into three physical qubits:

|0⟩_L = |000⟩

|1⟩_L = |111⟩

The encoding circuit for the bit-flip code can be implemented using two CNOT gates:

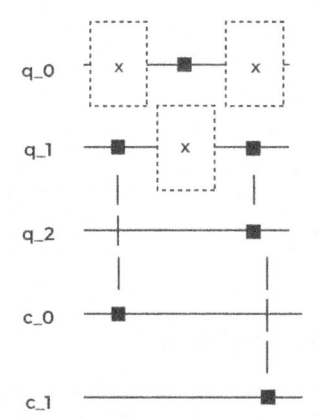

To detect and correct errors, the code uses two ancilla qubits (c_0 and c_1) to measure the parity of pairs of qubits (q_0 and q_1, q_1 and q_2). If a single bit-flip error occurs, the parity measurements will identify the flipped qubit, and a corrective operation can be applied to restore the original state.

Shor's Nine-Qubit Code:

Shor's nine-qubit code is a concatenated code that combines the three-qubit bit-flip code and the three-qubit phase-flip code to correct both types of single-qubit errors. The code encodes one logical qubit into nine physical qubits:

|0⟩_L = (|000⟩ + |111⟩) ⊗ (|000⟩ + |111⟩) ⊗ (|000⟩ + |111⟩) / 2√2

|1⟩_L = (|000⟩ - |111⟩) ⊗ (|000⟩ - |111⟩) ⊗ (|000⟩ - |111⟩) / 2√2

The encoding circuit for Shor's nine-qubit code involves applying the bit-flip encoding to each set of three qubits, followed by the phase-flip encoding to the three sets:

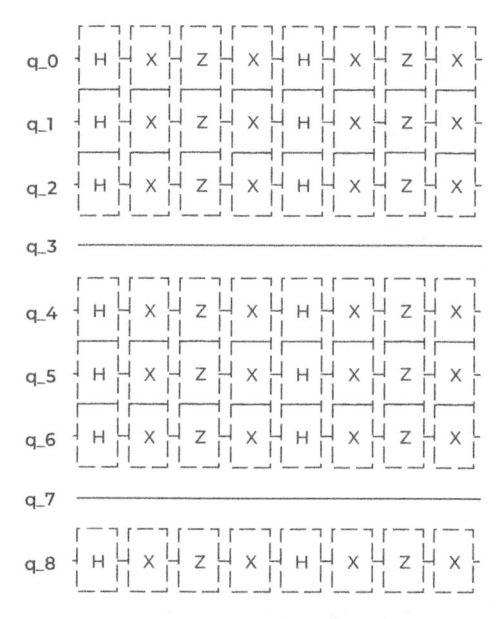

```
q_0 ─┤H├─┤X├─┤Z├─┤X├─┤H├─┤X├─┤Z├─┤X├─
q_1 ─┤H├─┤X├─┤Z├─┤X├─┤H├─┤X├─┤Z├─┤X├─
q_2 ─┤H├─┤X├─┤Z├─┤X├─┤H├─┤X├─┤Z├─┤X├─
q_3 ─────────────────────────────────
q_4 ─┤H├─┤X├─┤Z├─┤X├─┤H├─┤X├─┤Z├─┤X├─
q_5 ─┤H├─┤X├─┤Z├─┤X├─┤H├─┤X├─┤Z├─┤X├─
q_6 ─┤H├─┤X├─┤Z├─┤X├─┤H├─┤X├─┤Z├─┤X├─
q_7 ─────────────────────────────────
q_8 ─┤H├─┤X├─┤Z├─┤X├─┤H├─┤X├─┤Z├─┤X├─
```

The error correction procedure involves measuring the parity of pairs of qubits in each set of three qubits (bit-flip correction) and measuring the parity of pairs of sets (phase-flip correction). If an error is detected, a corrective operation is applied to the affected qubit or set.

Steane Code (Seven-Qubit Code):

The Steane code is a quantum error correction code that encodes one logical qubit into seven physical qubits. It is a CSS (Calderbank-Shor-Steane) code, which means that it uses separate encodings for bit-flip and phase-flip errors. The Steane code can correct arbitrary single-qubit errors.

The logical basis states of the Steane code are:

$|0\rangle\_L = (|0000000\rangle + |1010101\rangle + |0110011\rangle + |1100110\rangle + |0001111\rangle + |1011010\rangle + |0111100\rangle + |1101001\rangle) / \sqrt{8}$

$|1\rangle\_L = (|1111111\rangle + |0101010\rangle + |1001100\rangle + |0011001\rangle + |1110000\rangle + |0100101\rangle + |1000011\rangle + |0010110\rangle) / \sqrt{8}$

The encoding circuit for the Steane code is more complex than the previous examples and involves a series of CNOT and Hadamard gates:

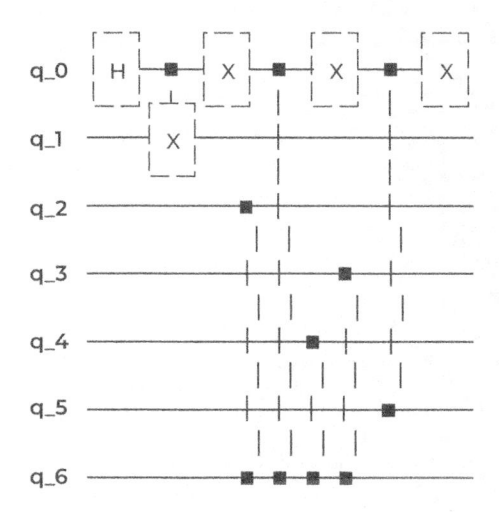

```
q_0  ─H─■─X─■─X─■─X─
        │   │   │
q_1  ───X───┼───┼───
            │   │
q_2  ───────■───┼───
            │   │
q_3  ───────┼───■───
            │   │
q_4  ───────■───┼───
                │
q_5  ───────────■───

q_6  ─────────■■■■──
```

The error correction procedure for the Steane code involves measuring a set of six stabilizer generators, which are operators that reveal the presence of errors without disturbing the encoded logical state. The stabilizer measurements are used to identify the type and location of the error, and a corrective operation is applied accordingly.

Surface Code

The surface code is a topological quantum error correction code that encodes logical qubits in a two-dimensional lattice of physical qubits. The surface code has a high threshold for error correction, meaning that it can tolerate a relatively high level of noise and imperfections in the physical qubits.

In the surface code, each logical qubit is encoded using a square lattice of physical qubits, with data qubits located on the vertices and ancilla qubits located on the edges. The stabilizer generators of the surface code are four-qubit operators that involve the data qubits and the neighboring ancilla qubits. The error correction procedure for the surface code involves measuring the stabilizer generators in a specific order and using the measurement outcomes to identify and correct errors. The surface code can correct both bit-flip and phase-flip errors, as well as certain types of correlated errors.

PRACTICAL CHALLENGES FOR REALIZING FAULT TOLERANCE

The stringent accuracy thresholds required to break even in correcting errors make demonstrating a fault-tolerant system extremely challenging. The complex nature of multi-quit measurements also introduces control complications. Significant hardware improvements will be essential to definitively surpass fault tolerance break-even points and maintain a quantum advantage.Reducing logical qubit error rates to orders of magnitude below the underlying physical error rates is a prerequisite for any quantum advantage. This demands high qubit fidelities, manufacturability, and robust classical control systems. Performing stabilizer measurements needed for large-scale codes on noisy hardware introduces additional challenges. Systematically characterizing error mechanisms and noise modes in different qubit modalities guide the development of tailored error mitigation techniques. While clear theoretical paths forward exist, realizing even small-scale fault tolerance on actual quantum computers remains one of the central outstanding engineering challenges in the field.

ERROR MITIGATION TECHNIQUES

Beyond quantum error correction, additional software techniques like artificially increasing measurement fidelity, dynamic decoupling and noise tailoring of Hamiltonians, redundant circuits, and classical outer loop error suppression help mitigate specific hardware noise channels and improve algorithm outcomes pre-fault-tolerance. Measurements can be repeated and statistically analyzed to improve fidelity beyond bare qubit readout accuracy. Dynamical decoupling pulse sequences between gate operations help maintain coherence and cancel noise. Compiling algorithms into hardware-efficient circuits reduces gate count and cumulative errors. Tailoring driving pulses to qubit device parameters compensates control crosstalk. Classically modeling the noise and then subtracting its estimated contribution filters errors. Such error mitigation strategies attack from all levels of the software stack to approximate error-free computation on noisy devices. While not replacements for error correction, they help bridge the gap and provide value in the NISQ era.

CHAPTER 4
QUANTUM ALGORITHMS AND QUANTUM SOFTWARE

Quantum Algorithm Design Principles: Complexity and Quantum Oracle

Quantum algorithms must be designed differently from classical algorithms due to the principles of quantum mechanics. Two key aspects enable quantum algorithms to outperform classical counterparts - the ability to represent and evaluate a function or problem in a quantum superposition and the use of a quantum oracle to query that function. This section explains these unique capabilities and their implications on the complexity and design of quantum algorithms.

LEVERAGING QUANTUM PARALLELISM AND SUPERPOSITION

Quantum superposition allows the representation of multiple inputs and outputs of a function simultaneously. This massive parallelism can be leveraged to evaluate a function using far fewer operations than classically needed. However, maintaining and using a superposition requires quantum resources like qubits. Designing algorithms to minimize qubit requirements is an important consideration.

Superposition enables massive parallelism by allowing a quantum state to encode exponentially many inputs at once. This means a function can be evaluated for all those inputs in one go. Classical algorithms lack this ability and must process each input sequentially. Superposition is what gives quantum algorithms their ability to provide speedups over classical approaches.

However, superpositions are delicate and require quantum bits or qubits to maintain. The number of qubits limits the size of the superposition and parallelism achievable. A key consideration in quantum algorithm design is minimizing qubit requirements while still representing the necessary inputs to achieve the desired speedup. There are algorithmic techniques like recursion and database compression that can reduce qubit costs.

QUERYING FUNCTIONS VIA QUANTUM ORACLES

While a superposition represents multiple function values in parallel, a quantum oracle provides the capability to query or evaluate the function on all those values simultaneously. This, again, provides a speedup over classical algorithms, which query functions sequentially. The design of an appropriate quantum oracle to match the problem is key.

Quantum oracles are black boxes that perform function evaluations over superpositions. This allows for assessing the function of many inputs in parallel. Classically, each input would have to be evaluated sequentially in separate function calls. The quantum oracle is what enables the efficient extraction of information - like finding periods or matches - from the superposition.

The design of the quantum oracle is problem-specific and crucial for the algorithm's performance. It needs to efficiently map the desired function into a reversible quantum operation. The oracle complexity relates to the number of queries needed. Optimized Oracle design reduces overall time complexity and can make previously intractable problems solvable.

EXTRACTING SOLUTIONS BY MANIPULATING SUPERPOSITIONS

Quantum algorithms manipulate a superposition of inputs via an oracle to arrive at a superposition of outputs that reveals the desired solution upon measurement. This allows solving problems faster, but the solution still needs to be extracted through repeated runs and measurements. Algorithm design focuses on maximizing the probability of measuring the solution.

The quantum algorithm applies operators like the oracle and quantum Fourier transform to manipulate the superposition. This extracts the hidden solutions into the phase or amplitudes of the output states. But the solution still needs to be measured, collapsing the superposition. Algorithms are designed to increase the likelihood of the collapsed state revealing the solution.

For example, Grover's algorithm amplifies the amplitude of solution states so they have a higher probability of being measured. Shor's algorithm uses quantum signal processing via the QFT to estimate periods from the output phase. The algorithms run multiple times to statistically sample the output superposition and reconstruct the complete solution.

ANALYZING COMPLEXITY: QUERY AND TIME COMPLEXITY

The complexity of quantum algorithms is assessed via query complexity - the number of oracle queries needed - and time complexity - the gate operations used. These capture the quantum speedup compared to optimal classical algorithms for a problem. Reducing query and operational complexity is key to efficient quantum algorithms.

Query complexity counts the Oracle function evaluations needed. This captures the power of the quantum oracle to assess inputs in superposition parallelly. Time complexity measures the gate operations by the algorithm around the oracle. Together, they assess the computational resources required and represent the quantum speedup.

For example, Grover's algorithm has O(1) query complexity versus O(N) classically but still has O(sqrt(N)) gate operations. Shor's factoring algorithm demonstrates both O(log N) query complexity and O((log N)3) time complexity - an exponential speedup over the best classical factoring methods.

NOTABLE QUANTUM ALGORITHM SPEEDUPS

Grover's algorithm for searching demonstrates quadratic speedup in query complexity over classical algorithms. Shor's algorithm demonstrates an exponential speedup in time complexity compared to the best-known classical factoring techniques.

Grover's O(sqrt(N)) search algorithm provides a quadratic speedup over the classical O(N) linear search but is optimal. Shor's polynomial time quantum factoring algorithm exponentially improves on the sub-exponential runtime of the best classical number field sieve algorithm.

These demonstrate the potential for quantum algorithms to provide both polynomial and exponential speedups over classical approaches for certain problems by leveraging quantum parallelism and quantum Fourier transforms. However, not all problems can achieve such speedups.

LIMITS TO QUANTUM SPEEDUPS

While quantum algorithms can offer speedups, not all problems lend themselves to this quantum advantage. Algorithm design relies on matching the right problem structure to quantum principles to gain an advantage. Quantum algorithms provide limited speedup for problems requiring brute force search or evaluation.

Quantum algorithms excel where quantum resources like superposition and entanglement can be leveraged. But problems requiring brute force evaluation on all inputs may, at best, have polynomial speedups. Additionally, the advantages diminish if the output requires processing an exponential number of bits.

For example, while Grover's algorithm provides a square root speedup, it still searches classically. Using quantum walks and other techniques can improve the speedups possible via enhanced quantum approaches tailored to the problem structure.

QUANTUM WALK-BASED ALGORITHMS

Quantum walks allow the designing of quantum algorithms with superior spatial and time complexity compared to classical random walks. Quantum walk-based algorithms have been developed for element distinctness, graph traversal, and other problems providing quadratic or exponential speedups. Quantum walks harness quantum interference to traverse graphs or data structures faster than classical random walks. Algorithms based on quantum walk principles can detect unique elements in a list, traverse graphs, and search databases with demonstrable speedups over classical methods. For example, quantum walks can traverse a glued tree graph exponentially faster in O(sqrt(N)) time versus O(N) classically. Quantum walk search achieves up to quadratic speedup over Grover's algorithm for structured problems like two-dimensional grid traversal.

APPLICATION OF AMPLITUDE AMPLIFICATION

Amplitude amplification, rooted in principles from Grover's algorithm, offers a generalized technique for quadratic improvements in sampling and optimization tasks. It enhances the probability of measuring a desired solution from a superposition, providing a quadratic speedup over classical algorithms.

This technique proves beneficial in various applications, such as Monte Carlo integration, optimization tasks, and statistical analysis of graph properties. Amplitude amplification also elevates algorithms for solving linear systems, phase estimation, and differential equations.

In-depth Exploration of Shor's Algorithm for Prime Factorization

Shor's factoring algorithm, devised by Peter Shor in 1994, provides exponential speedup over the best-known classical factoring techniques. It can find the prime factors of large integers in polynomial time, a task considered classically intractable. This potentially breaks widely used RSA encryption. We discuss Shor's algorithm in detail, including the quantum principles it exploits, and walk through an example.

LEVERAGING QUANTUM PARALLELISM VIA PERIOD FINDING

Shor's algorithm relies on using a period-finding subroutine to find the order or period of a chosen function. This period can then be used to factor the integer. It essentially reduces factoring to finding periods, a problem well suited for quantum computation.

Shor's algorithm transforms the factoring problem into finding the period of a function. It exploits the power of quantum parallelism to efficiently find periods, even for functions that are difficult to analyze classically. Factoring is then just a classical post-processing step using the period.

This is achieved by choosing a suitable quantum function like modular exponentiation and putting it in a superposition over all inputs. The period of this function corresponds directly to the factors of the input number. Estimating this quantum period provides an exponential advantage.

APPLYING THE QUANTUM FOURIER TRANSFORM

The quantum part of Shor's algorithm puts the function, here modular exponentiation, in a superposition, allowing its period to be extracted efficiently. Classically, this would require an infeasible number of evaluations to find periodicity.

The quantum Fourier transform enables extracting the period of the function from its superposition. It takes advantage of interference effects to essentially find the period in only one shot, unlike the exponential time needed classically.

In essence, modular exponentiation "kicks" the period into the phase of the superposition states. The quantum Fourier transform then amplifies this phase information into the measured amplitudes, from which the period can be estimated. This provides the exponential speedup.

COMBINING QUANTUM AND CLASSICAL STEPS

After the quantum period finding routine, the period is used classically to find the factors via continued fractions or greatest common divisors. The combination of quantum and classical steps enables an efficient factorization.

The period extracted through the quantum component provides the necessary insight to factorize the number using classical math techniques. Different strategies like Euclid's algorithm or the continuous fraction method can turn the period into prime factors.

The power of Shor's algorithm comes from using a quantum subroutine to efficiently obtain the period information, which would be unavailable to classical algorithms in polynomial time. The classically intractable factorization problem becomes solvable by using quantum resources to transform its structure.

PROVING THE QUANTUM ADVANTAGE

Shor's algorithm demonstrates that factoring is in the complexity class BQP - bounded error, quantum, and polynomial time. It can factor an n-bit number using $O((\log n)3)$ operations, unlike sub-exponential classical algorithms.

Shor's algorithm theoretically proves factoring can be performed on a quantum computer in polynomial time with bounded error. No classical algorithm can factor integers in better than sub-exponential time. This demonstrates definitively the exponential quantum speedup.

The poly-logarithmic time complexity of $O((\log N)3)$ operations for Shor's algorithm versus the best classical algorithm with $O(\exp((64/9)1/3 (\log N)1/3))$ complexity highlights the immense advantage provided by quantum computation.

IMPLEMENTING MODULAR EXPONENTIATION

The quantum circuit for Shor's algorithm applies modular exponentiation as the function oracle to a uniform superposition over all input states created using Hadamard gates. This "kicks" the period information into the phase, which is then extracted using the quantum Fourier transform.

Modular exponentiation maps neatly to a reversible quantum circuit. The modular exponential function can be calculated using modular multiplication in log time. This allows efficient implementation of the oracle to apply it over the uniform superposition of inputs.

Running this circuit over the superposed inputs creates output states encoding the period in the relative phases. The quantum

Fourier transform then reveals this hidden period information to exponential speedup over corresponding classical algorithms.

ANALYZING QUANTUM SPEEDUP FACTORS

Quantum parallelism enables the simultaneous evaluation of the function for all superposition states. The quantum Fourier transform leverages interference effects to efficiently identify the period in a single iteration, resulting in an exponential advantage over classical factoring approaches.

Quantum parallelism allows for the assessment of the function over all exponentially many superposed inputs in one go. Additionally, the quantum Fourier transform capitalizes on interference to extract the period from this extensive processing. In contrast, classical methods necessitate separate evaluations for each input, resulting in exponential time complexity even with optimization and parallelization. The combination of quantum parallelism and Fourier signal processing thus delivers exponential speedup.

IMPLEMENTATION CHALLENGES

Shor's algorithm does have high qubit and circuit depth requirements, which are currently limiting practical implementations. But even small demonstrations conclusively prove the ability to factor integers in polynomial time beyond classical capability.

The number of qubits grows linearly with the number being factored. Factoring 1024-bit numbers requires thousands of logical qubits. The modular exponentiation also requires deep circuits. These make scaling Shor's algorithm challenging.

However, small-scale demonstrations have factored numbers like 15 and 21, conclusively proving the concept and viability of the approach. Even such small implementations are impossible for classical computers, demonstrating clear quantum advantage.

IMPLICATIONS FOR RSA ENCRYPTION

Shor's algorithm carries profound implications for RSA encryption, widely employed in securing internet communications. RSA's security relies on the classical difficulty of factoring large numbers, a foundation that Shor's algorithm challenges. Scalable implementations of Shor's algorithm could potentially compromise RSA encryption.

RSA encryption, built on the classical difficulty of factoring large numbers, faces vulnerability in a world with large quantum computers. Quantum computers could employ a scalable implementation of Shor's algorithm to break RSA encryption, prevalent in most secure communication protocols on the internet today. This necessitates focused research into new quantum-resistant cryptographic schemes.

DEVELOPING QUANTUM-SAFE CRYPTOGRAPHY

Post-quantum cryptography is being developed to provide security in a world with large quantum computers. New cryptosystems like lattice-based encryption are believed to be quantum resilient, unlike RSA. But guaranteeing quantum-proof security remains an open challenge.

Cryptosystems like NTRU, BLISS, and SPHINCS+ based on lattices rather than factoring are considered secure against attacks from a quantum computer running Shor's algorithm.

However, proving any cryptosystem is unconditionally secure against both classical and quantum attacks is mathematically challenging. Work continues on designing encryption schemes with provable quantum security and efficient implementations.

Example

Let's walk through Shor's algorithm for factoring 15 using the function $f(x) = a^x \mod 15$. We choose $a=7$ and put $f(x)$ in superposition via Hadamard gates. The period finding extracts $r=4$, leading to factors 3 and 5.

Consider the function $f(x) = 7^x \mod 15$ for factoring 15. The circuit implements this function in superposition. Applying the QFT reveals a period $r=4$. By picking $7^x=1 \mod 15$, we get $7^4=1 \mod 15$, giving factors 3 and 5.

This small example illustrates the working of Shor's algorithm. The same approach is applied for larger numbers, where the period finding via superposition and QFT provides exponential speedup over classical factoring techniques.

PRACTICAL IMPLEMENTATION CONSIDERATIONS

Executing Shor's algorithm requires choosing the right modular exponentiation function and a number of qubits to represent the number being factored and function outputs. Small examples can be run on quantum programming platforms like IBM Quantum for experimenting.

Factors like the number of qubits, circuit depth, gate errors, runtime, and post-processing overheads need consideration for practical implementation. Small instances can be simulated classically or run on quantum hardware like IBM Quantum systems.

Quantum assembly languages like Qiskit simplify programming Shor's algorithm. Modular functions and circuit optimizations can be explored to improve performance and noise resilience when demonstrating larger number factoring.

QUANTUM ERROR CORRECTION INTEGRATIONS

Integrating Shor's algorithm with quantum error correction will be essential for realizing large-scale fault-tolerant implementations. This requires developing optimized circuits with lower qubit overheads and space-time complexity compared to naively integrating error correction.

Efficient error-corrected modular multiplication circuits and fault-tolerant non-clifford gates are active areas of research. Reducing qubit and gate counts via algorithm-specific optimizations instead of generic error correction will facilitate scalability.

RESOURCE OPTIMIZED CIRCUITS

Circuit optimizations aim to minimize the qubit and gate complexity of modular exponentiation and the quantum Fourier transform. Techniques like time-space resource optimization, Multi-Qubit gates, and carry look-ahead adders are being investigated to realize more compact quantum circuits.

Hardware-efficient circuit implementations will help scale Shor's algorithm to larger numbers by reducing quantum memory and operation requirements. This builds on existing modular multiplication optimizations but focuses on depth-space complexity trades.

APPLICATION SPECIFIC CUSTOMIZATION

Research also focuses on customizing and extending Shor's algorithm for specific application domains. These include factoring elliptic curve products, quantum simulations, improving security in cryptographic implementations, and quantum chemistry simulations.

Application-specific adaptations allow further extraction advantage by optimizing Shor's algorithm routines and circuits around the problem structure. Focus areas include reducing overhead in elements like function choice, problem encoding, and measurement.

QUANTUM-CLASSICAL HYBRID APPROACHES

Hybrid quantum-classical variants of Shor's algorithm aim to deliver benefits before fully fault-tolerant devices are available. For instance, performing modular multiplication with a shallow quantum circuit and measuring before the QFT provides partial information to guide classical factoring.

Such co-processing and outsourcing approaches distribute the computation across available quantum and classical resources. Hybrid algorithms expand the reach of NISQ hardware to deliver useful quantum advantages for practical applications.

ANALYZING IMPLEMENTATION TRADEOFFS

Analysis of space-time complexity tradeoffs for Shor's algorithm implementation will guide design choices for specific hardware. Time-optimal implementations may require more qubits, while space-optimal designs increase circuit depth.

Understanding these implementation tradeoff spaces is key to mapping Shor's algorithm onto quantum accelerators with limited qubit numbers or coherence times. This requires analyzing modular exponentiation in terms of both circuit depth and width.

Grover's Algorithm: Optimized Search in Unsorted Databases

Grover's quantum search algorithm provides quadratic speedup for searching unsorted databases over classical algorithms by utilizing amplitude amplification principles. We discuss how it achieves the speedup compared to classical search along with complexity analysis and applications. An illustrated example explains Grover's algorithm.

QUADRATIC SPEEDUP OVER CLASSICAL SEARCH

Classically, searching an unsorted database of N entries requires O(N) queries in the worst case to find a matching entry. Grover's algorithm can find the match in just O(sqrt(N)) queries, providing quadratic speedup.

Sequential search algorithms check each element one by one, resulting in linear time complexity. Grover's quantum algorithm accesses all elements in superposition via an oracle query and amplifies the solution through inversion about the mean for quadratic speedup.

This demonstrates how quantum algorithms can leverage principles like superposition, interference, and amplitude amplification to provide polynomial speedups over classical approaches for certain problems.

MINIMIZING QUERY COMPLEXITY

Grover's algorithm provides this quadratic speedup in terms of query complexity - the number of Oracle function evaluations needed. This captures the power of assessing the oracle over superpositions.

Query complexity specifically measures the number of times the oracle function needs to be evaluated on different inputs. This reflects how efficiently the search space can be processed. Grover's O(sqrt(N)) query complexity demonstrates a square root speedup over the classical O(N) complexity.

While the gate complexity remains comparable, the quantum algorithm offers an exponential advantage in parallelizing the brute force search through querying in superposition.

OPTIMAL QUANTUM SEARCH

Grover's algorithm is optimal in query complexity for unstructured database search. It provides a quadratic improvement that is theoretically the maximum possible for quantum search without additional problem structure.

Information-theoretic proofs establish lower bounds, showing that Grover's O(sqrt(N)) scaling cannot be improved for general black-box search. This optimality relies on maintaining a uniform superposition without prior knowledge of solution distribution.

However, utilizing problem structure through tailored wavefunction evolutions or quantum walk-based techniques can provide beyond quadratic speedups. Grover's algorithm sets a baseline demonstrating quantum search advantage.

APPLICATIONS OF GROVER'S ALGORITHM

Some key applications of Grover's algorithm include searching molecular databases for drug discovery, cryptographic code-breaking, and searching data structures like classification trees and unsorted tables.

Database search tasks are directly accelerated using Grover's algorithm. Other applications like DNA sequence alignment, pattern matching, and decoding encrypted data benefit from quadratic speedups in the search routines.

The algorithm is also applicable to navigating graphs and networks, optimization problems, and machine learning algorithms relying on search over unordered spaces. Grover's algorithm provides a versatile quantum subroutine.

GENERALIZATION VIA AMPLITUDE AMPLIFICATION

The principles of amplitude amplification, initially employed in Grover's algorithm for search, have been extended to yield quadratic speedups in diverse applications, including sampling, counting, and finding minimums from an unsorted set. This extension underscores the broad utility of amplitude amplification.

The core technique of selectively amplifying target states proves versatile, offering quadratic speedups in areas such as resource estimation, statistical analysis, and optimization. These applications leverage the fundamental principle of Grover's algorithm - the strategic evolution of a superposition to enhance the probability of measuring desired states. The application of amplitude amplification extends the quantum advantage into previously unexplored domains.

HARDWARE IMPLEMENTATION CONSIDERATIONS

In terms of hardware, Grover's algorithm exhibits low qubit requirements but significant circuit depth due to its iterative structure. Practical implementations address this challenge by decomposing the algorithm into optimized block-wise circuits, aligning with constraints on coherence times.

While the quadratic scaling in query complexity maintains low qubit costs, the deep circuits resulting from iterative amplification pose challenges for near-term hardware. Segmenting the circuits becomes imperative to minimize superposition loss before completing the algorithm. Reducing circuit depth through parallelization, custom gates, and optimized arithmetic blocks tailored for amplitude amplification is essential for the successful implementation of emerging quantum hardware.

Illustrative Examples and Experiments

An illustrative example involves applying Grover's algorithm to find a marked item in an unordered list of 16 entries, requiring merely 4 iterations or "queries," as opposed to the classical necessity of 16.

This example highlights the scalability of the number of iterations, which scales as the square root of the number of entries (N). Experiments using quantum programming platforms like IBM Qiskit can validate the quadratic speedup over classical brute force search, even for small lists. Code samples are available for implementing Grover's algorithm in search tasks, enabling both classical simulation and practical execution on quantum hardware for performance analysis.

Extensions and Variants

Beyond its foundational form, extensions of Grover's algorithm explore multi-solution searching, dynamic databases, and synergies with other quantum algorithms. Research endeavors include efforts to reduce circuit depth and enhance fault tolerance.

Enhancements feature variable phase shifts for multiple solutions, nested searches over structured databases, and hybrid schemes involving pre-processing with quantum Fourier transforms.

Ongoing investigations delve into fault-tolerance techniques employing codeword-stabilized ancillas. Overall, adapting Grover's algorithm to leverage problem structure and combining it with quantum subroutines to solve complex tasks remains an active area of research. Extensions aim to expand applicability.

Future Research Directions

Continued analysis and innovation in quantum search algorithms build on Grover's foundational approach to achieve beyond quadratic speedups. Novel techniques tailored to problem structure and optimized implementation will expand quantum search capabilities.

Possible directions include developing dedicated hardware accelerators optimized for search, designing error-corrected implementations, and using Grover as a subroutine in larger quantum algorithms. Discovering new structured search problems amenable to quantum speedups is also key.

PART 2

Mastering Quantum Computing Techniques

2

CHAPTER 5

BUILDING QUANTUM CIRCUITS

Quantum Circuit Design Principles and Circuit Composition

Quantum circuit design demands a solid understanding of the foundational principles governing qubit behaviour, gate operations, and measurement. Additionally, the composition of modular circuits using gate sequences, subcircuits, and libraries facilitates clear and optimized implementations. This section comprehensively covers core quantum circuit design concepts and methods for constructing composite circuits from basic components.

QUBIT INITIALIZATION

Qubit initialization is the process of setting a qubit (quantum bit) into a desired initial state before performing quantum operations. In quantum computing, qubits are the fundamental units of information, analogous to classical bits in conventional computing. However, unlike classical bits, which can only be in one of two states (0 or 1), qubits can be in a superposition of multiple states simultaneously.

The most common qubit states used for initialization are:

1. $|0\rangle$ (ket 0): This represents the ground state or the computational basis state of 0.
2. $|1\rangle$ (ket 1): This represents the excited state or the computational basis state of 1.
3. $|+\rangle$ (ket plus): This represents an equal superposition of $|0\rangle$ and $|1\rangle$ states, i.e., $(|0\rangle + |1\rangle)/\sqrt{2}$.
4. $|-\rangle$ (ket minus): This represents an equal superposition of $|0\rangle$ and $|1\rangle$ states with a relative phase of -1, i.e., $(|0\rangle - |1\rangle)/\sqrt{2}$.

To initialize a qubit, various techniques can be employed, depending on the physical implementation of the qubit. Some common methods include:

1. Cooling: Qubits can be initialized to the ground state ($|0\rangle$) by cooling them to extremely low temperatures, close to absolute zero, using techniques like dilution refrigeration.
2. Optical pumping: In some qubit implementations, such as trapped ions or neutral atoms, laser pulses can be used to initialize the qubits to a desired state by selectively exciting or de-exciting specific energy levels.
3. Microwave pulses: Applying microwave pulses of specific frequencies and durations can be used to initialize qubits, particularly in superconducting qubit systems.
4. Measurement and feedback: Measuring the state of a qubit and applying corrective pulses based on the measurement outcome can help initialize the qubit to a desired state.

Accurate and reliable qubit initialization is crucial for performing quantum algorithms and achieving high-fidelity quantum operations. Imperfections in qubit initialization can lead to errors and decoherence, limiting the accuracy and reliability of quantum computations.

I can provide a textual representation of qubit initialization using quantum circuit notation. In this notation, qubits are represented as horizontal lines, and quantum gates are shown as boxes or symbols on these lines.

Initializing a qubit to the $|0\rangle$ state:

$|0\rangle$ ─────────

Initializing a qubit to the $|1\rangle$ state:

$|0\rangle$ ───────── X

Here, the X represents the Pauli-X gate, which is equivalent to a classical NOT gate, flipping the state from $|0\rangle$ to $|1\rangle$.

Initializing a qubit to the $|+\rangle$ state:

$|0\rangle$ ───────── H

The H represents the Hadamard gate, which creates an equal superposition of the $|0\rangle$ and $|1\rangle$ states.

Initializing a qubit to the $|-\rangle$ state:

$|0\rangle$ ───────── H ───────── Z

The H represents the Hadamard gate, and the Z represents the Pauli-Z gate, which introduces a relative phase of -1 between the $|0\rangle$ and $|1\rangle$ states.

These are simplified representations of qubit initialization in a quantum circuit.

QUANTUM GATES

Quantum gates are the fundamental building blocks of quantum circuits, analogous to classical logic gates in conventional computers. They are unitary operations that manipulate the state of one or more qubits. Quantum gates can be represented as matrices, and applying a gate to a qubit is equivalent to multiplying the qubit's state vector by the corresponding matrix.

Some of the most commonly used quantum gates include:

Pauli Gates:

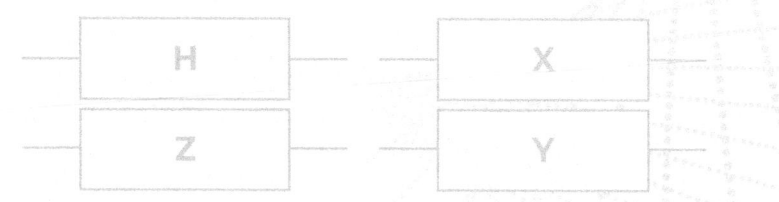

─[H]─   ─[X]─

─[Z]─   ─[Y]─

- Pauli-X (X) Gate: Equivalent to a classical NOT gate, it flips the state of a qubit from |0⟩ to |1⟩ and vice versa.
- Pauli-Y (Y) Gate: Applies a π rotation around the Y-axis of the Bloch sphere.
- Pauli-Z (Z) Gate: Applies a π rotation around the Z-axis of the Bloch sphere, introducing a relative phase of -1 between the |0⟩ and |1⟩ states.

Hadamard (H) Gate:

―――| H |―――

- Creates an equal superposition of the |0⟩ and |1⟩ states. It is often used to initialize qubits and create entanglement.

Phase Gates:
- Phase (S) Gate: Applies a π/2 phase shift to the |1⟩ state.
- T Gate: Applies a π/4 phase shift to the |1⟩ state.

Rotation Gates:
- Rotation X (RX) Gate: Performs a rotation around the X-axis of the Bloch sphere by a specified angle.
- Rotation Y (RY) Gate: Performs a rotation around the Y-axis of the Bloch sphere by a specified angle.
- Rotation Z (RZ) Gate: Performs a rotation around the Z-axis of the Bloch sphere by a specified angle.

Controlled Gates:
- Controlled-NOT (CNOT or CX) Gate: A two-qubit gate that flips the state of the target qubit if the control qubit is in the |1⟩ state.
- Controlled-Z (CZ) Gate: A two-qubit gate that applies a phase of -1 to the target qubit if both the control and target qubits are in the |1⟩ state.
- Controlled-Phase (CPhase) Gate: A two-qubit gate that applies a specified phase to the target qubit if both the control and target qubits are in the |1⟩ state.

Swap Gate:
- Exchanges the states of two qubits.

Toffoli (CCNOT) Gate:

$$CCNOT = \begin{pmatrix} 1 & 0 & 0 & 0 & 0 & 0 & 0 & 0 \\ 0 & 1 & 0 & 0 & 0 & 0 & 0 & 0 \\ 0 & 0 & 1 & 0 & 0 & 0 & 0 & 0 \\ 0 & 0 & 0 & 1 & 0 & 0 & 0 & 0 \\ 0 & 0 & 0 & 0 & 1 & 0 & 0 & 0 \\ 0 & 0 & 0 & 0 & 0 & 1 & 0 & 0 \\ 0 & 0 & 0 & 0 & 0 & 0 & 0 & 1 \\ 0 & 0 & 0 & 0 & 0 & 0 & 1 & 0 \end{pmatrix}$$

- A three-qubit gate that flips the state of the target qubit if both control qubits are in the |1⟩ state.

Fredkin (CSWAP) Gate:
- A three-qubit gate that swaps the states of the second and third qubits if the first qubit (control) is in the |1⟩ state.

These are just a few examples of the many quantum gates used in quantum computing. Quantum circuits are built by applying a

$$\begin{pmatrix} 1 & 0 & 0 & 0 & 0 & 0 & 0 & 0 \\ 0 & 1 & 0 & 0 & 0 & 0 & 0 & 0 \\ 0 & 0 & 1 & 0 & 0 & 0 & 0 & 0 \\ 0 & 0 & 0 & 1 & 0 & 0 & 0 & 0 \\ 0 & 0 & 0 & 0 & 1 & 0 & 0 & 0 \\ 0 & 0 & 0 & 0 & 0 & 0 & 1 & 0 \\ 0 & 0 & 0 & 0 & 0 & 1 & 0 & 0 \\ 0 & 0 & 0 & 0 & 0 & 0 & 1 & 0 \end{pmatrix}$$

sequence of quantum gates to qubits, enabling the implementation of quantum algorithms and computations. The specific set of quantum gates used depends on the quantum hardware and the algorithm being implemented.

GATE RULES AND LIMITATIONS

Quantum gates adhere to rules derived from matrix mathematics and quantum mechanics. Physically realizable gates must be unitary, preserving total probability. Gate combinations must avoid unwanted phase kickbacks. Adhering to established gate constraints enables feasible circuit design.

Quantum gates represent valid physical operations on qubits, with unitary constraints conserving probability amplitudes during state evolution. Non-unitary gates would enable impossible operations like qubit cloning. Phase kickback describes how conditional gates can accidentally imprint phase factors on unaffected qubits, introducing errors.

Understanding gate limitations allows designers to strategize optimizations and circuit architectures respecting the laws of quantum mechanics. Methodical matrix calculations ensure unitarity. Quantum error-correcting codes counteract phase kickback and other noise processes to improve fidelity. Overall, quantum gates provide powerful tools when applied within their mathematical frameworks.

MEASUREMENT

Qubit measurement collapses superposition states into observable basis states. The measurement basis determines the possible measurement outcomes. Intermediate measurements during algorithm execution require resetting the qubit state afterwards.

Measurement extracts classical information from qubits by projecting quantum states into defined bases like the Z basis $\{|0\rangle, |1\rangle\}$. The measurement result is probabilistic, with outcome likelihoods determined by qubit amplitudes. Repeated measurements sample the underlying quantum distribution. Intermediate measurements within circuits destroy existing superpositions.

Qubit measurement enables a classical readout of quantum computation results. The chosen measurement basis should match the target algorithm observables. Readout fidelity limits discrimination between close states. Partial collapsing measurements provide some information while maintaining coherence. Architecting circuits around measurement processes is key, from readout error correction to resetting post-measurement.

REVERSIBLE LOGIC

Reversible logic is a fundamental concept in quantum computing, as all quantum operations must be reversible to maintain the coherence and unitary evolution of the quantum state. In classical computing, many logical operations are irreversible, meaning that the input cannot be uniquely determined from the output. For example, the AND operation takes two input bits and produces one output bit, but given the output, it is impossible to determine the unique input combination that produced it.

In quantum circuits, all gates must be reversible, meaning that the input state can be uniquely determined from the output state. This is achieved through the use of unitary operations, which preserve the norm and orthogonality of the quantum state vector.

Example 1: NOT Gate (Pauli-X Gate)

The NOT gate, also known as the Pauli-X gate, is a simple example of a reversible operation. It flips the state of a qubit from $|0\rangle$ to $|1\rangle$ and vice versa:

$|0\rangle$ ———— X ———— $|1\rangle$

$|1\rangle$ ———— X ———— $|0\rangle$

The NOT gate is its own inverse, meaning that applying it twice in succession returns the qubit to its original state:

$|0\rangle$ ———— X ———— X ———— $|0\rangle$

$|1\rangle$ ———— X ———— X ———— $|1\rangle$

Example 2: Controlled-NOT (CNOT) Gate

The CNOT gate is a two-qubit reversible gate that flips the state of the target qubit if the control qubit is in the $|1\rangle$ state:

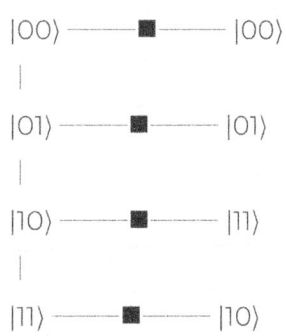

$|00\rangle$ ————■———— $|00\rangle$
|
$|01\rangle$ ————■———— $|01\rangle$
|
$|10\rangle$ ————■———— $|11\rangle$
|
$|11\rangle$ ————■———— $|10\rangle$

The CNOT gate is also its own inverse, and applying it twice in succession returns the qubits to their original states.

Example 3: Toffoli (CCNOT) Gate The

Toffoli gate, also known as the CCNOT gate, is a three-qubit reversible gate that flips the state of the target qubit if both control qubits are in the $|1\rangle$ state:

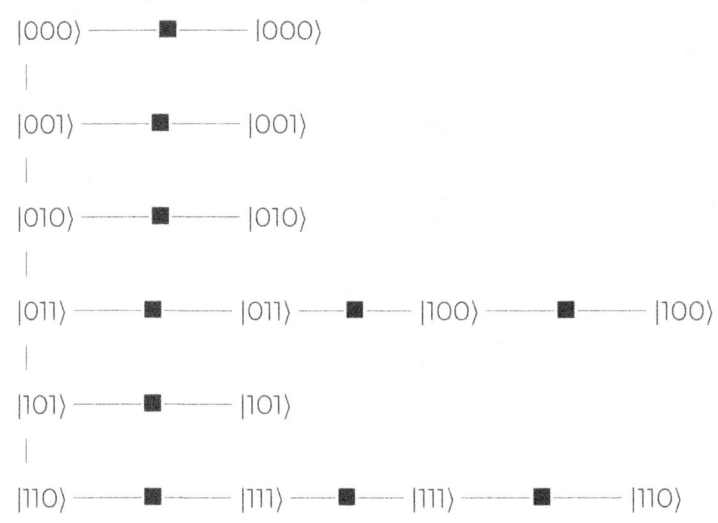

$|000\rangle$ ———■——— $|000\rangle$

$|001\rangle$ ———■——— $|001\rangle$

$|010\rangle$ ———■——— $|010\rangle$

$|011\rangle$ ———■——— $|011\rangle$ ———■——— $|100\rangle$ ———■——— $|100\rangle$

$|101\rangle$ ———■——— $|101\rangle$

$|110\rangle$ ———■——— $|111\rangle$ ———■——— $|111\rangle$ ———■——— $|110\rangle$

The Toffoli gate is universal for classical reversible computation, meaning that any reversible classical circuit can be constructed using only Toffoli gates.

Ancilla Qubits and Uncomputation

To implement complex reversible operations, quantum circuits often employ ancilla qubits to store intermediate results and maintain reversibility. However, these ancilla qubits can become entangled with the main qubits, leading to unwanted "garbage" outputs. To remove these garbage outputs and reset the ancilla qubits, quantum circuits use a technique called uncomputation.

Example 4: Uncomputation

Consider a quantum circuit that performs a complex operation on a qubit, using an ancilla qubit to store an intermediate result:

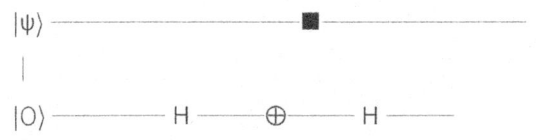

$|\psi\rangle$ ————————————■————————————

$|0\rangle$ ———— H ———— ⊕ ———— H ————

After the operation, the ancilla qubit is entangled with the main qubit, forming a garbage output. To remove this garbage output, we can apply the inverse of the operation (uncomputation) to disentangle the ancilla qubit:

$|\psi\rangle$ ———————■———————————————■———

$|0\rangle$ ——— H ——⊕—— H ——…—— H ——⊕—— H ——

By applying the inverse operation, the ancilla qubit is returned to its original $|0\rangle$ state, and the garbage output is removed.

Reversible logic is a crucial concept in quantum circuit design, as it ensures the coherence and unitary evolution of the quantum state. By using reversible gates, ancilla qubits, and uncomputation, quantum circuits can implement complex operations while maintaining the reversibility required for quantum computing.

GATE LIBRARIES

Gate libraries are collections of quantum gates that can be used to construct quantum circuits. These libraries provide a set of basic building blocks that can be combined to implement more complex operations and algorithms. Gate libraries are essential for designing and optimizing quantum circuits, as they abstract away the low-level details of the underlying quantum hardware and provide a higher-level interface for quantum programming.

Example 1: IBM Qiskit Gate Library

IBM's Qiskit is a popular open-source quantum computing framework that provides a comprehensive gate library. Some of the basic gates in Qiskit include:

1. Single-qubit gates:
   - Pauli gates (X, Y, Z)
   - Hadamard gate (H)
   - Rotation gates (RX, RY, RZ)
   - Phase gates (S, T)
2. Two-qubit gates:
   - Controlled-NOT (CNOT or CX)
   - Controlled-Z (CZ)

- Swap gate

3. Three-qubit gates:
   - Toffoli gate (CCNOT)

Example Qiskit code for applying gates:

pythonCopy codefrom qiskit import QuantumCircuit

qc = QuantumCircuit(2) # Create a quantum circuit with 2 qubits

qc.h(0) # Apply Hadamard gate to qubit 0

qc.cx(0, 1) # Apply CNOT gate with qubit 0 as control and qubit 1 as target

qc.rz(0.5, 1) # Apply RZ gate with angle 0.5 to qubit 1

Example 2: Google Cirq Gate Library

Google's Cirq is another open-source quantum computing framework with its own gate library. Some of the basic gates in Cirq include:

1. Single-qubit gates:
   - Pauli gates (X, Y, Z)
   - Hadamard gate (H)
   - Rotation gates (Rx, Ry, Rz)
   - Phase gates (S, T)

2. Two-qubit gates:
   - Controlled-NOT (CNOT)
   - Controlled-Z (CZ)
   - Swap gate
   - Sqrt(SWAP) gate

Example Cirq code for applying gates:

pythonCopy codeimport cirq

qubits = cirq.LineQubit.range(2) # Create 2 qubits

circuit = cirq.Circuit(

*   cirq.H(qubits[0]), # Apply Hadamard gate to qubit 0*

*   cirq.CNOT(qubits[0], qubits[1]), # Apply CNOT gate with qubit 0 as control and qubit 1 as target*

*   cirq.Rz(0.5)(qubits[1]) # Apply Rz gate with angle 0.5 to qubit 1*

)

In addition to these basic gates, quantum computing frameworks often provide higher-level operations and algorithms built on top of the gate libraries. These include quantum Fourier transforms, amplitude amplification, and quantum error correction circuits. Gate libraries also play a crucial role in quantum circuit optimization and compilation. By providing a standardized set of gates, these libraries enable the development of optimization techniques that can reduce the gate count and depth of quantum circuits, making them more efficient and less prone to errors. As quantum hardware evolves, gate libraries are also expanding to include new types of gates and operations that are tailored to specific quantum architectures. For example, some quantum processors may support additional gates or have different connectivity constraints, requiring the gate libraries to adapt accordingly.

SUBCIRCUITS

Subcircuits, also known as subroutines or modules, are reusable components of a quantum circuit that perform a specific task or implement a particular quantum operation. By breaking down a complex quantum circuit into smaller, modular subcircuits, designers can make the overall circuit more manageable, easier to understand, and less prone to errors. Subcircuits also promote code reuse and make it easier to update and maintain quantum circuits.

Example 1: Quantum Adder Subcircuit

A quantum adder is a subcircuit that performs the addition of two quantum registers. It takes two input registers (A and B) and an ancilla qubit (carry) and stores the result of the addition in one of the input registers (A).

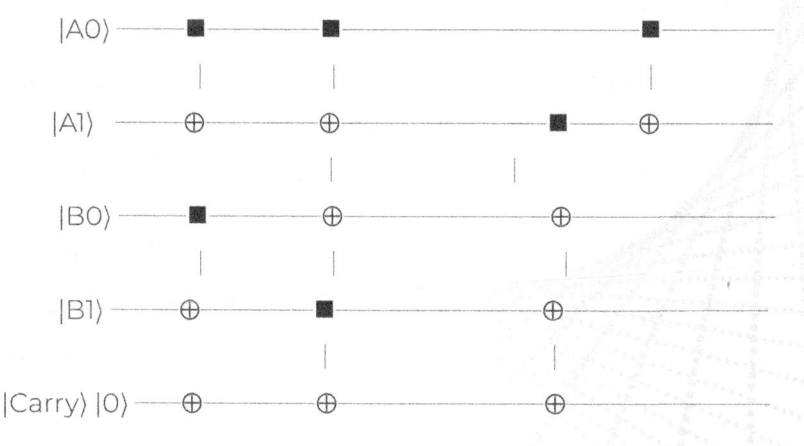

This quantum adder subcircuit can be used as a building block for more complex arithmetic operations, such as quantum multiplication or division.

Example 2: Quantum Fourier Transform (QFT) Subcircuit

The quantum Fourier transform (QFT) is a fundamental subcircuit used in many quantum algorithms, such as Shor's algorithm for factoring and the phase estimation algorithm. The QFT performs a Fourier transform on the amplitudes of a quantum state, mapping it from the computational basis to the Fourier basis.

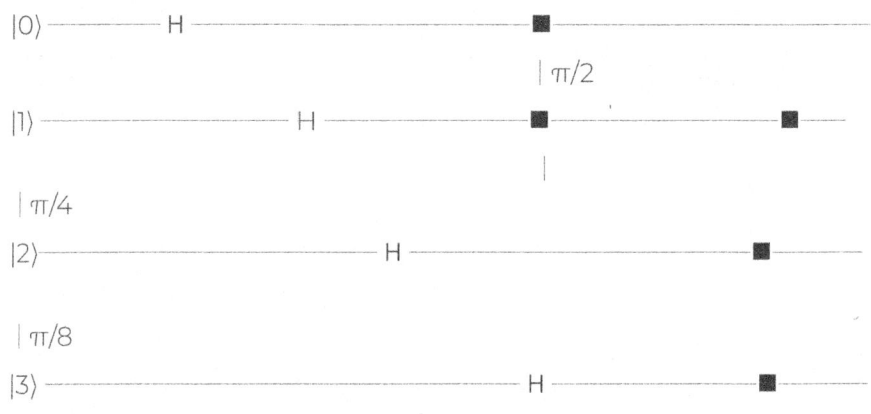

```
|0⟩ ———— H ————————————————————■————————————————
                              | π/2
|1⟩ ———————————— H ————————————■————————————■———
                                             |
| π/4                                        |
|2⟩ ———————————————————— H ——————————————————■——
                                              
| π/8
|3⟩ ———————————————————————————— H ———————————■—
```

The QFT subcircuit consists of a series of Hadamard gates and controlled-phase rotations, with the phase rotations depending on the position of the qubit in the register.

Example 3: Quantum Error Correction Subcircuits

Quantum error correction is essential for mitigating the effects of noise and errors in quantum circuits. Error correction subcircuits encode logical qubits into larger entangled states, allowing for the detection and correction of errors.

One example of an error correction subcircuit is the three-qubit bit-flip code:

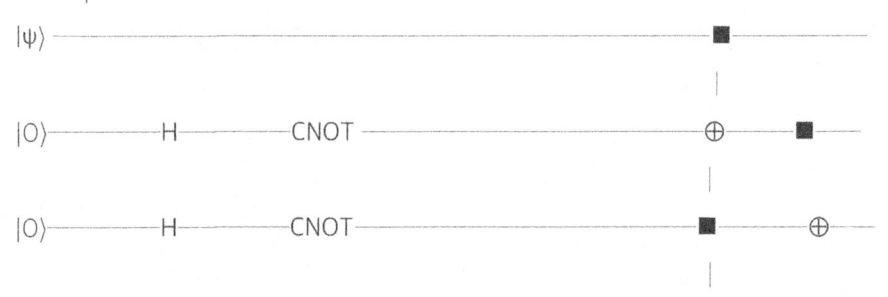

```
|ψ⟩ ——————————————————————————————————————■————————
                                          |
|0⟩ ———— H ———— CNOT ——————————————————— ⊕ ——■———
                                              |
|0⟩ ———— H ———— CNOT ——————————————————— ■ ——⊕———
```

```
                    ■----■----■
                    |    |    |
                    |    |    H----
                    |
                    H--------------
```

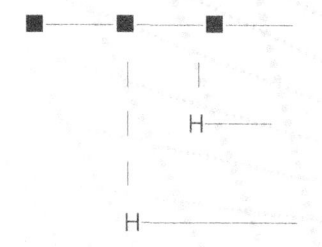

This subcircuit encodes a single logical qubit into three physical qubits, allowing for the detection and correction of single-qubit bit-flip errors.

By using subcircuits, quantum circuit designers can create modular, reusable components that make it easier to build and analyze complex quantum algorithms. Subcircuits also provide a way to encapsulate quantum operations and error correction techniques, making quantum circuits more robust and scalable.

MAPPING TO QUANTUM HARDWARE

Mapping a quantum circuit to quantum hardware involves translating the abstract gates and qubits of the circuit into the physical operations and qubits of a specific quantum processor. This process is crucial for executing quantum circuits on real quantum devices and is often constrained by the limitations and characteristics of the underlying hardware.

Example: Mapping a Quantum Circuit to IBM's Qiskit Hardware

Consider the following quantum circuit:

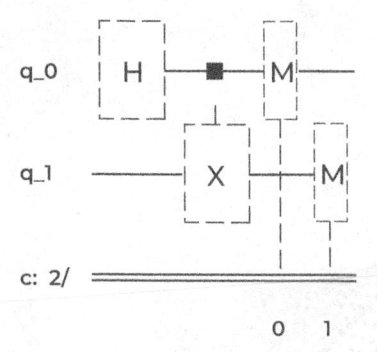

```
q_0    ─┤ H ├──■──┤ M ├───
                │   └─┬─┘
q_1    ────────┤ X ├─┤ M ├
                └───┘ └─┬─┘
c: 2/  ════════════════╩═╩═
                       0 1
```

To map this circuit to IBM's Qiskit hardware, we need to consider the following:

1. Qubit connectivity: IBM's quantum processors have limited connectivity between qubits. For example, the IBM Q 5 Yorktown device has the following connectivity:

```
0 -- 1 -- 2
     |    |
     3    4
```

If the qubits in the circuit are not directly connected, we may need to insert additional SWAP gates to move the quantum states around.

2. Gate decomposition: Some gates in the circuit may not be directly supported by the hardware. In this case, we need to decompose these gates into a sequence of native gates that the hardware can execute. For example, the Hadamard gate (H) might be decomposed into a sequence of rotation gates (Rx and Rz) and a phase gate (S).

3. Timing and pulse control: The hardware may have specific timing constraints and pulse control requirements for executing gates and measurements. The mapping process needs to take these into account and ensure that the circuit can be executed within the available time slots and with the appropriate pulse sequences.

Here's an example of how the above circuit might be mapped to the IBM Q 5 Yorktown device using Qiskit:

```python
from qiskit import QuantumCircuit, execute, IBMQ

# load IBMQ account and backend
IBMQ.load_account()
provider = IBMQ.get_provider('ibm-q')
backend = provider.get_backend('ibmq_qasm_simulator')

# Create a quantum circuit
qc = QuantumCircuit(2, 2)

# Apply gates
qc.h(0)
qc.cx(0, 1)
qc.measure([0, 1], [0, 1])
```

```
# Map the circuit to the backend
mapped_circuit = transpile(qc, backend)

# Execute the circuit on the backend
result = execute(mapped_circuit, backend, shots=1024).result()
counts = result.get_counts(mapped_circuit)
print(counts)
```

In this example, the **transpile** function is used to map the circuit to the backend, taking into account the connectivity constraints and decomposing gates as needed. The resulting **mapped_circuit** is then executed on the backend using the **execute** function.

CIRCUIT REPRESENTATION

Quantum circuits can be represented in various ways, both for visualization and for programming. Some common representations include:

1. Quantum circuit diagrams: These are graphical representations of quantum circuits, using symbols for qubits, gates, and measurements. Examples include the circuit diagram shown earlier, created using ASCII art.

2. Matrix representations: Quantum gates and circuits can be represented as matrices acting on the quantum state vector. For example, the Hadamard gate (H) can be represented as: $$H = \frac{1}{\sqrt{2}}\begin{bmatrix} 1 & 1 \\ 1 & -1 \end{bmatrix}$$

3. Quantum assembly language (QASM): QASM is a low-level language for describing quantum circuits, similar to assembly language for classical computers. OpenQASM is an example of a QASM language used by IBM's Qiskit framework.

4. Quantum programming languages: High-level programming languages, such as IBM's Qiskit, Google's Cirq, and Microsoft's Q#, provide abstractions for creating and manipulating quantum circuits using familiar programming constructs like functions, classes, and control flow statements.

5. Quantum intermediate representation (IR): Quantum IRs are lower-level representations of quantum circuits, used for optimization and compilation. Examples include the QIR (Quantum Intermediate Representation) used by Microsoft's Q# compiler

and the QIL (Quantum Instruction Language) used by the Liquid framework.

These different representations serve various purposes, from visualizing and designing circuits to implementing and executing them on quantum hardware and simulators. The choice of representation depends on the specific requirements of the task, the level of abstraction needed, and the tools and frameworks being used.

DESIGN TOOLS

Several design tools are available for creating, simulating, and analyzing quantum circuits. These tools range from low-level quantum assembly languages to high-level quantum programming languages and graphical user interfaces (GUIs). Some popular quantum circuit design tools include:

IBM Qiskit

Qiskit is an open-source quantum computing framework that provides tools for creating, manipulating, and visualizing quantum circuits. It includes a high-level quantum programming language (Qiskit Terra), a quantum simulator (Qiskit Aer), and tools for accessing IBM's quantum hardware (Qiskit IBMQ Provider). Qiskit also has a GUI called Qiskit Composer for designing circuits visually.

Example: Creating a quantum circuit to implement the Bell state

from qiskit import QuantumCircuit, execute, IBMQ, Aer

 # Create a quantum circuit with 2 qubits and 2 classical bits
qc = QuantumCircuit(2, 2)

 # Apply a Hadamard gate to the first qubit
qc.h(0)

 # Apply a CNOT gate with the first qubit as the control and the second qubit as the target
qc.cx(0, 1)

```
# Measure both qubits and store the results in the classical bits
qc.measure([0, 1], [0, 1])

# Simulate the circuit using the Qiskit Aer simulator
backend = Aer.get_backend('qasm_simulator')
result = execute(qc, backend, shots=1024).result()
counts = result.get_counts(qc)
print(counts)
```

This example demonstrates how to create a quantum circuit using Qiskit, apply gates (Hadamard and CNOT), measure the qubits, and simulate the circuit using the Qiskit Aer simulator.

Google Cirq

Cirq is an open-source quantum programming framework developed by Google. It provides a Python API for creating and manipulating quantum circuits, as well as tools for simulating and running circuits on Google's quantum hardware.

Example: Creating a quantum circuit to implement the Deutsch-Jozsa algorithm

```
import cirq

# Define the oracle function
def oracle(qubits):
    # Implement the oracle function here
    # ...

# Create a quantum circuit
qubits = cirq.LineQubit.range(3)
circuit = cirq.Circuit()

# Apply Hadamard gates to all qubits
circuit.append(cirq.H.on_each(qubits))
```

```python
# Apply the oracle function
circuit.append(oracle(qubits))

# Apply Hadamard gates to the first two qubits
circuit.append(cirq.H.on_each(qubits[:2]))

# Measure the first two qubits
circuit.append(cirq.measure(qubits[:2]))

# Simulate the circuit
simulator = cirq.Simulator()
result = simulator.run(circuit, repetitions=100)
print(result)
```

This example shows how to create a quantum circuit using Cirq, define an oracle function, apply gates (Hadamard), measure the qubits, and simulate the circuit using the Cirq simulator.

Microsoft Q#

Q# is a domain-specific programming language for quantum computing developed by Microsoft. It integrates with the .NET framework and provides tools for creating, simulating, and running quantum circuits on Microsoft's quantum simulator (QDK) and quantum hardware (Azure Quantum).

Example: Creating a quantum circuit to implement the Grover's search algorithm

```
using Microsoft.Quantum.Canon;
using Microsoft.Quantum.Intrinsic;

operation OracleFunction(markedElements : Int[], reg : Qubit[]) : Unit {
    // Implement the oracle function here
    // ...
}
```

```
operation GroverSearch(markedElements : Int[]) : Int {
    // Determine the number of qubits needed
    let nQubits = markedElements.Length;

    // Allocate qubits
    use qubits = Qubit[nQubits];

    // Apply Hadamard gates to all qubits
    ApplyToEach(H, qubits);

    // Perform Grover's iterations
    for (_ in 1..NIterations(nQubits)) {
        OracleFunction(markedElements, qubits);
        ApplyGroverDiffusion(qubits);
    }

    // Measure the qubits
    let result = MeasureInteger(LittleEndian(qubits));

    // Return the measurement result
    return result;
}
```

This example demonstrates how to create a quantum circuit using Q#, define an oracle function, apply gates (Hadamard), perform Grover's iterations, measure the qubits, and return the result.

Rigetti PyQuil

PyQuil is a Python library for creating and running quantum circuits on Rigetti's quantum hardware and simulator. It includes a quantum instruction language called Quil, which is used to describe quantum circuits and programs.

Amazon Braket

Amazon Braket is a fully managed quantum computing service

that provides access to quantum hardware and simulators from various providers, including D-Wave, IonQ, and Rigetti. It includes a Python SDK for creating and running quantum circuits, as well as a GUI for designing circuits visually.

Quantum Development Kit (QDK):

The QDK is a set of tools for developing quantum applications in Q#, including a quantum simulator, a quantum computer tracer, and extensions for Visual Studio and Jupyter Notebooks.

Quirk

Quirk is a browser-based quantum circuit simulator that provides a drag-and-drop interface for creating and simulating quantum circuits. It is designed for educational purposes and allows users to easily visualize the state of a quantum circuit at each step of its execution.

These tools provide different levels of abstraction and cater to different use cases, from research and experimentation to education and application development. They enable users to design, simulate, and run quantum circuits on various platforms, from local simulators to cloud-based quantum hardware.

DESIGN METHODOLOGY

Methodical design flows involving abstractions, decomposition, modules, and layers translate algorithms into optimized gate sequences. Rigorous design, verification, and validation are essential for building functional circuits.

Top-down design, involving steps like specification, high-level design, optimization, simulation, prototyping, and testing, refines concepts into detailed implementations. Careful abstraction management avoids mismatches between layers, and design automation gains importance for complex chips.

As quantum circuit design matures, adopting structured design practices becomes necessary to manage growing complexity, akin to classical computing. This transition opens opportunities to adapt classical techniques like hardware-software co-design to meet quantum's unique requirements.

HYBRID QUANTUM-CLASSICAL

Hybrid quantum-classical systems combine quantum and classical computing to leverage the strengths of both paradigms. These systems typically involve a quantum processor working in conjunction with a classical computer, where the classical computer is used for tasks such as control, optimization, and error correction.

Example: Variational Quantum Eigensolver (VQE)

- VQE is a hybrid quantum-classical algorithm used for solving eigenvalue problems, particularly in the context of quantum chemistry.
- The quantum processor is used to prepare a parameterized quantum state and measure the expectation value of the Hamiltonian operator.
- The classical computer is used to optimize the parameters of the quantum state to minimize the expectation value of the Hamiltonian.
- The process is iterative, with the classical computer updating the parameters based on the measurement results from the quantum processor until convergence is achieved.

import numpy as np

from qiskit import QuantumCircuit, execute, Aer

def create_ansatz_circuit(params):

  qc = QuantumCircuit(2)

  qc.ry(params[0], 0)

  qc.ry(params[1], 1)

  qc.cx(0, 1)

  return qc

def measure_hamiltonian(qc, hamiltonian):

  backend = Aer.get_backend('qasm_simulator')

  result = execute(qc, backend).result()

  counts = result.get_counts(qc)

  *energy = sum(hamiltonian[b] * v for b, v in counts.items()) / su-*

```
m(counts.values())
    return energy


def vqe(hamiltonian, params, num_iterations):
    for _ in range(num_iterations):
        qc = create_ansatz_circuit(params)
        energy = measure_hamiltonian(qc, hamiltonian)
        params = optimize_parameters(params, energy)
    return params, energy


# Example usage
hamiltonian = {'00': -1, '11': 1}
initial_params = [0.0, 0.0]
num_iterations = 100
optimized_params, minimum_energy = vqe(hamiltonian, initial_params, num_iterations)
print(f"Optimized parameters: {optimized_params}")
print(f"Minimum energy: {minimum_energy}")
```

In this example, the **create_ansatz_circuit** function creates a parameterized quantum circuit (ansatz) based on the given parameters. The **measure_hamiltonian** function executes the quantum circuit and measures the expectation value of the Hamiltonian. The **vqe** function performs the iterative optimization process, where the classical computer updates the parameters to minimize the energy.

AUTOMATED DESIGN

Automated design refers to the use of computational tools and algorithms to assist in the design and optimization of quantum circuits. The goal is to automate certain aspects of the quantum circuit design process, making it more efficient and less prone to human error.

Example: Quantum Circuit Optimization using Genetic Algorithms

- Genetic algorithms (GAs) are a class of optimization algorithms inspired by the principles of natural selection and evolution.
- In the context of quantum circuit design, GAs can be used to optimize the parameters of a quantum circuit to minimize a specific cost function, such as the error rate or the gate count.
- The GA maintains a population of candidate solutions (quantum circuits), each represented by a set of parameters.
- The fitness of each candidate solution is evaluated based on the cost function.
- The GA applies genetic operators, such as selection, crossover, and mutation, to evolve the population over multiple generations.
- The best solution found by the GA represents the optimized quantum circuit.

```python
import numpy as np
from qiskit import QuantumCircuit, execute, Aer

def create_circuit(params):
    qc = QuantumCircuit(2)
    qc.rx(params[0], 0)
    qc.ry(params[1], 1)
    qc.cx(0, 1)
    qc.measure_all()
    return qc

def evaluate_fitness(qc, target_state):
    backend = Aer.get_backend('qasm_simulator')
    result = execute(qc, backend).result()
    counts = result.get_counts(qc)
    fidelity = counts.get(target_state, 0) / sum(counts.values())
    return fidelity

def genetic_algorithm(population_size, num_generations, target_state):
```

QUANTUM COMPUTING

```python
    population = np.random.rand(population_size, 2) * 2 * np.pi
    for _ in range(num_generations):
        fitness_scores = [evaluate_fitness(create_circuit(params), target_state) for params in population]
        parents = select_parents(population, fitness_scores)
        offspring = crossover_and_mutate(parents)
        population = offspring
    best_solution = population[np.argmax(fitness_scores)]
    return best_solution

# Example usage
population_size = 20
num_generations = 50
target_state = '11'
optimized_params = genetic_algorithm(population_size, num_generations, target_state)
optimized_circuit = create_circuit(optimized_params)
print(f"Optimized parameters: {optimized_params}")
print("Optimized circuit:")
print(optimized_circuit)
```

In this example, the create_circuit function creates a quantum circuit based on the given parameters. The evaluate_fitness function evaluates the fitness of a candidate solution by measuring the fidelity of the circuit's output with respect to the target state. The genetic_algorithm function implements the GA optimization process, evolving the population of candidate solutions over multiple generations.

DESIGN VERIFICATION

Verifying circuit implementations match specifications requires simulation, formal methods, testing, debuggers, and instrumentation. This ensures correctness before costly hardware execution.

Verification is challenged by state space explosion exacerbating classical techniques. Hybrid quantum-classical, modular, and approximate solutions ameliorate complexity. Equivalence checking confirms that the implementation matches the specification or abstract model. Verification tool integration into quantum software frameworks provides seamless design flows.

Rigorous verification supports engineering quality quantum circuits as complexity scales up. Beyond functionality, verification ensures hardware compatibility, resilience to noise, algorithmic optimality, security properties, and resource constraints are satisfied. Verification gaps could delay delivering fault-tolerant, practical quantum computing.

DESIGN GOALS AND TRADEOFFS

Quantum circuit designers must balance competing objectives, including minimization of gate counts, circuit depth, qubit usage, SWAPs, ancillas, noise, and errors. This complex optimization landscape necessitates design tradeoffs.

Improving one metric often worsens another. Lower circuit depth reduces noisy evolution but can require more qubits. Ancilla minimization conflicts with constructing reversible logic. Architecting circuits involves manoeuvring through subtle tradespaces via approximations, heuristics, and hardware awareness.

Quantum circuit design fundamentally grapples with resource limitations and noisy environments. No perfect solutions exist. Instead, designers judiciously navigate design goals prioritization based on application needs, hardware maturity, and practical constraints. Developing broad intuition for navigating these multidimensional tradespaces will be learned over time and experience.

Optimizing Quantum Gates: Unitary Operators and Gate Decomposition

Quantum algorithms require many gates for realization. Optimal gate counts reduce errors and decoherence while minimizing execution time on quantum hardware. Gate optimization techniques include employing efficient unitary operators and decomposing complex gates into simpler primitives.

UNITARY OPERATORS

Unitary operators are fundamental to quantum computing, as they describe the evolution of quantum states and the operations performed by quantum gates. A unitary operator U is a complex square matrix that satisfies the condition $U^\dagger U = U U^\dagger = I$, where $U^\dagger$ is the conjugate transpose of U, and I is the identity matrix.

Properties of unitary operators:

- Unitary operators preserve the norm of the quantum state vector, ensuring that the total probability of all possible outcomes is always 1.
- Unitary operators are reversible, meaning that for every unitary operator U, there exists an inverse operator $U^\dagger$ such that $U^\dagger U = U U^\dagger = I$.
- Unitary operators can be composed to create more complex quantum gates and circuits.

Example: Hadamard Gate

The Hadamard gate (H) is a single-qubit unitary operator that creates an equal superposition of the |0⟩ and |1⟩ states. Its matrix representation is:

$H = 1/\sqrt{2} * [1\ 1]$

$[1\ -1]$

Applying the Hadamard gate to the |0⟩ state:

$H |0\rangle = 1/\sqrt{2} * (|0\rangle + |1\rangle)$

This operation transforms the qubit from the |0⟩ state to an equal superposition of |0⟩ and |1⟩.

Example: Controlled-NOT (CNOT) Gate

The CNOT gate is a two-qubit unitary operator that performs a NOT operation on the target qubit if the control qubit is in the |1⟩ state. Its matrix representation is:

CNOT = [1 0 0 0]

[0 1 0 0]

[0 0 0 1]

[0 0 1 0]

Applying the CNOT gate to the state $|10\rangle$:

CNOT $|10\rangle = |11\rangle$

This operation flips the state of the target qubit (second qubit) from $|0\rangle$ to $|1\rangle$ because the control qubit (first qubit) is in the $|1\rangle$ state.

Unitary operators are essential for designing and optimizing quantum circuits, as they provide a mathematical framework for describing quantum gates and their effects on quantum states.

GATE COUNT REDUCTION

Gate count reduction is an important optimization technique that aims to minimize the number of gates in a quantum circuit while preserving its functionality. Reducing the gate count is crucial for several reasons:

- Quantum gates are prone to errors, and the more gates in a circuit, the higher the likelihood of errors accumulating.
- Quantum hardware has limited coherence times, meaning that qubits can only maintain their quantum state for a short duration. Minimizing the gate count helps to execute the circuit within the available coherence time.
- Reducing the gate count can lead to faster execution times and lower resource requirements, making the quantum algorithm more practical and scalable.

Example: Quantum Teleportation Circuit

Consider the quantum teleportation circuit, which allows the transfer of a quantum state from one qubit to another using entanglement and classical communication. A naive implementation of the circuit may look like this:

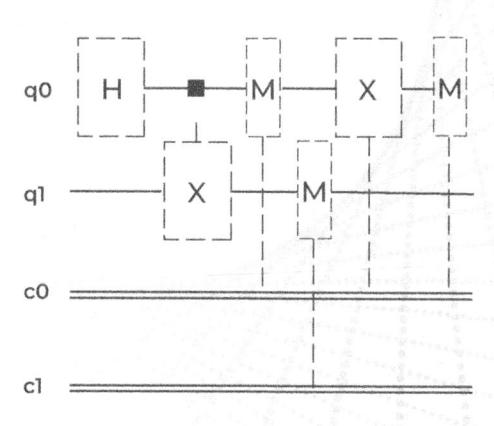

QUANTUM COMPUTING   121

In this circuit, the Hadamard gate (H) and the CNOT gate (X) are used to create entanglement between qubits q0 and q1. The measurement operations (M) and classical control operations (lines connecting measurements to gates) are used to transfer the state from q0 to q1.

Optimized circuit:

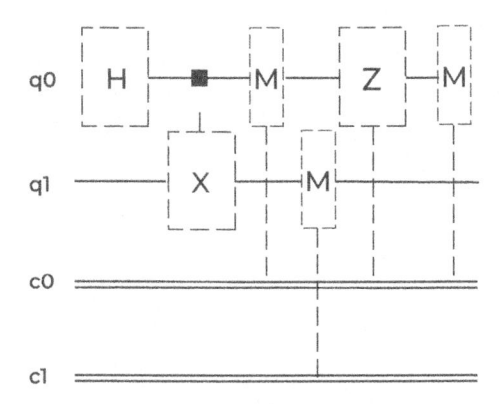

```
q0 ─| H |──■──|M|──| Z |──|M|
           │   │
q1 ────────| X |──|M|────────

c0 ═══════════════════════════

c1 ═══════════════════════════
```

In the optimized circuit, the second CNOT gate is replaced with a Z gate, which is a single-qubit operation. This optimization reduces the gate count by one, while still achieving the same functionality of transferring the quantum state from q0 to q1.

Example: Quantum Fourier Transform (QFT) Circuit

The quantum Fourier transform (QFT) is a key component in many quantum algorithms, such as Shor's algorithm for factoring and the phase estimation algorithm. A naive implementation of the QFT circuit for 4 qubits may look like this:

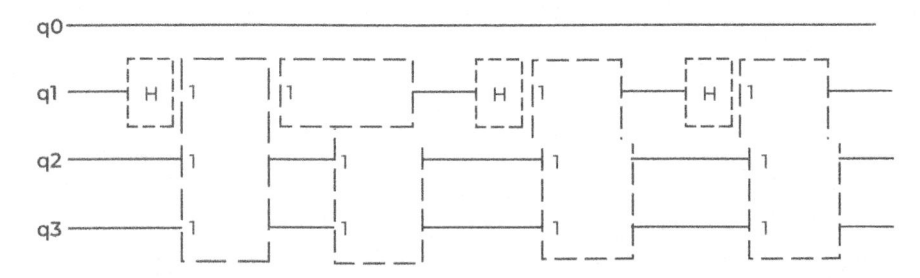

```
q0 ─────────────────────────────────────────────

q1 ─| H |─1────────1─| H |─1──────| H |─1───────

q2 ──────1─────────────────1──────────1─────────

q3 ──────────1─────────1─────────1──────────────
```

In this circuit, the Hadamard gates (H) and the controlled-phase rotation gates (1, representing a phase rotation of $\pi/2^k$, where k is the number of qubits between the control and target) are used to implement the QFT.

Optimized circuit:

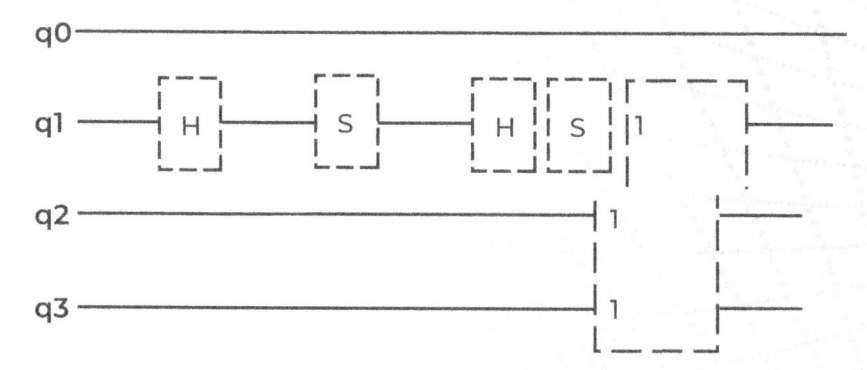

```
q0 ─────────────────────────────────────────
q1 ───[H]─────[S]─────[H][S][1]─────────────
q2 ──────────────────────────[1]────────────
q3 ──────────────────────────[1]────────────
```

In the optimized circuit, some of the controlled-phase rotation gates are replaced with single-qubit phase gates (S, representing a phase rotation of $\pi/2$). This optimization reduces the gate count and depth of the circuit, while still implementing the QFT.

GATE DECOMPOSITION

Gate decomposition is a technique used to break down complex quantum gates into a sequence of simpler, more fundamental gates. This is particularly useful when the target quantum hardware does not directly support the complex gate or when the complex gate is prone to errors. By decomposing the complex gate into a sequence of simpler gates, we can improve the reliability and efficiency of the quantum circuit.

Example: Toffoli Gate Decomposition

The Toffoli gate (CCNOT) is a three-qubit gate that performs a controlled-controlled-NOT operation. It can be decomposed into a sequence of single-qubit and two-qubit gates as follows:

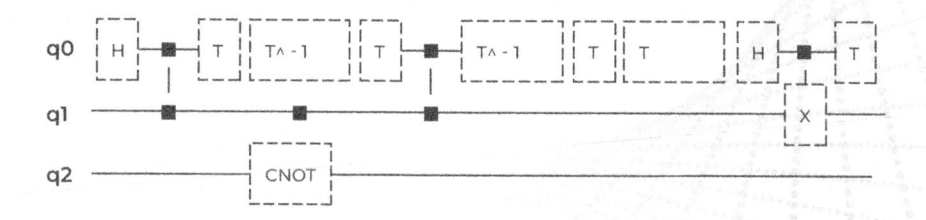

```
q0 ─[H]─●─[T]─[T^-1]─[T]─●─[T^-1]─[T]─[T]─[H]─●─[T]─
        │                │                    │
q1 ─────●────────────────●────────────────────[X]───
                                              
q2 ──────────[CNOT]─────────────────────────────────
```

In this decomposition, the Toffoli gate is replaced by a sequence of Hadamard (H), T, T^-1 (inverse of T), and CNOT gates. This decomposition allows the Toffoli gate to be implemented using only single-qubit and two-qubit gates, which are more commonly supported by quantum hardware.

Example: Controlled-Phase Gate Decomposition

The controlled-phase gate (CPhase) is a two-qubit gate that applies a phase rotation to the target qubit if the control qubit is in the |1⟩ state. It can be decomposed into a sequence of single-qubit and two-qubit gates as follows:

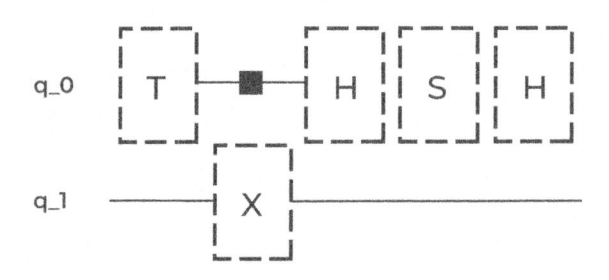

```
q_0  ─[T]─■─[H]─[S]─[H]─

q_1  ────[X]────────────
```

In this decomposition, the controlled-phase gate is replaced by a sequence of T, Hadamard (H), S, and CNOT (X) gates. This decomposition allows the controlled-phase gate to be implemented using only single-qubit and two-qubit gates, which are more commonly supported by quantum hardware.

Gate decomposition is an important technique in quantum circuit optimization, as it allows complex gates to be replaced by simpler, more reliable gates. By decomposing complex gates, we can also reduce the gate count and depth of the circuit, which can help to mitigate the effects of errors and improve the overall performance of the quantum algorithm.

COMMUTATIVITY

Commutativity is a property of quantum gates that refers to the ability to swap the order of two gates without changing the overall effect on the quantum state. When two gates commute, applying them in either order produces the same result. Commutativity is an important concept in quantum circuit optimization, as it allows for the reordering of gates to reduce the gate count, depth, or complexity of the circuit.

Example: Commutativity of Single-Qubit Gates

Consider the following single-qubit gates:

- Pauli-X (X) gate: Performs a bit-flip operation
- Pauli-Z (Z) gate: Performs a phase-flip operation

These gates commute with each other, meaning that applying them in either order produces the same result:

$X \cdot Z = Z \cdot X$

This property allows us to reorder the X and Z gates in a quantum circuit without changing the overall functionality.

Example: Commutativity of Controlled Gates

Consider the following two-qubit gates:

- Controlled-X (CX) gate: Performs a controlled-NOT operation
- Controlled-Z (CZ) gate: Performs a controlled-phase operation

When the control qubit is the same for both gates, they commute with each other:

$CX \cdot CZ = CZ \cdot CX$

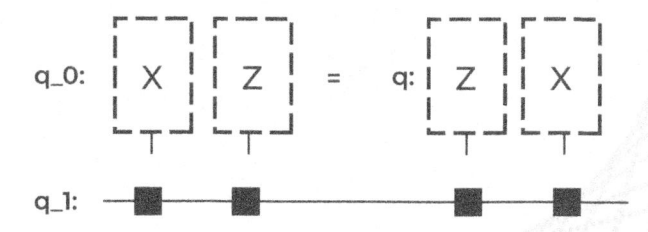

This property allows us to reorder the CX and CZ gates in a quantum circuit without changing the overall functionality, as long as the control qubit is the same for both gates.

QUANTUM COMPUTING

Example: Commutativity in Quantum Phase Estimation

In the quantum phase estimation algorithm, the commutativity of controlled-phase gates is exploited to simplify the circuit. The algorithm uses a series of controlled-phase gates with different phase rotations to extract the eigenvalue of a unitary operator.

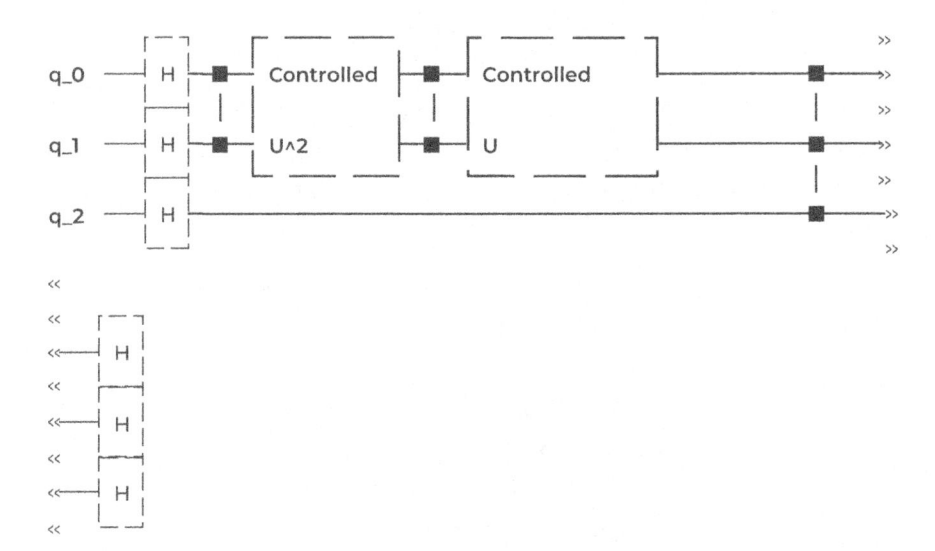

```
q_0 ──H──■──Controlled──■──Controlled──────────■──»
         │              │                      │  »
q_1 ──H──■──U^2─────────■──U───────────────────■──»
                                                  »
q_2 ──H───────────────────────────────────────────■──»
```

```
«
«  ┌─┐
«──┤H├
«  └─┘
«  ┌─┐
«──┤H├
«  └─┘
«  ┌─┐
«──┤H├
«  └─┘
«
```

In this circuit, the controlled-U and controlled-U^2 gates commute with each other, as they have the same control qubit (q_0) and act on different target qubits. This commutativity allows for the reordering of these gates, which can help to simplify the circuit and reduce the gate count.

CIRCUIT DEPTH REDUCTION

Circuit depth reduction is a technique used to minimize the number of time steps (depth) required to execute a quantum circuit. The depth of a quantum circuit is determined by the longest path from the input qubits to the output qubits, where each gate is considered to take one time step. Reducing the circuit depth is important for several reasons:

•Quantum hardware has limited coherence times, and shorter circuit depths help to ensure that the computation is completed before the qubits lose their quantum state due to decoherence.

- Shorter circuit depths can reduce the accumulation of errors, as fewer gates are applied to the qubits.
- Reducing the circuit depth can lead to faster execution times and improved performance of the quantum algorithm.

Example: Parallel Execution of Gates

One way to reduce the circuit depth is to identify gates that can be executed in parallel. Consider the following quantum circuit:

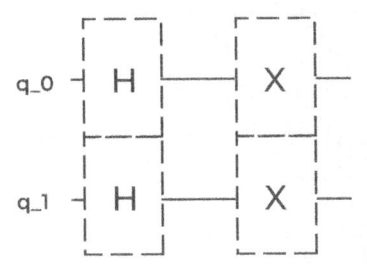

In this circuit, the Hadamard (H) gates can be executed in parallel on both qubits, followed by the Pauli-X (X) gates. By recognizing this parallelism, we can reduce the circuit depth from 4 to 2:

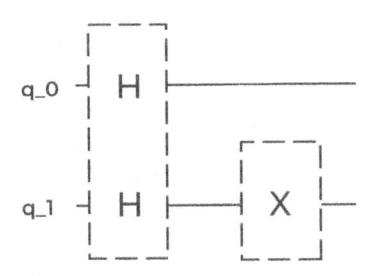

Example: Gate Commutation

Another technique for reducing circuit depth is to exploit the commutativity of gates. By reordering commuting gates, we can often reduce the depth of the circuit. Consider the following quantum circuit:

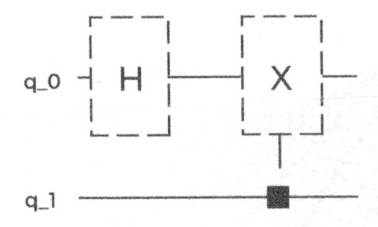

In this circuit, the Hadamard (H) gate on q0 commutes with the controlled-X (CX) gate. By swapping the order of these gates, we can reduce the circuit depth from 3 to 2:

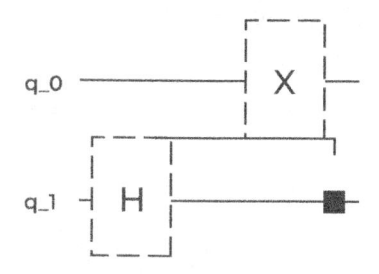

```
q_0 ─────────────┤ X ├──
                 └─┬─┘
      ┌───┐        │
q_1 ──┤ H ├────────■───
      └───┘
```

Circuit depth reduction is an important optimization technique that can help to create more efficient and error-resilient quantum circuits. By identifying opportunities for parallel execution and exploiting gate commutativity, we can minimize the depth of the circuit and improve its performance on quantum hardware.

REVERSIBLE LOGIC SYNTHESIS:

Reversible logic synthesis is the process of designing reversible logic circuits, which are circuits where the inputs can be uniquely determined from the outputs. In quantum computing, reversible logic is essential because all quantum gates are inherently reversible. The goal of reversible logic synthesis is to create efficient and optimized reversible circuits that implement the desired functionality while minimizing the number of gates, qubits, and circuit depth.

Example: Reversible Full Adder

A full adder is a digital circuit that adds three one-bit numbers (two operands and a carry-in) and produces a sum and a carry-out. To create a reversible full adder, we need to ensure that the circuit has the same number of inputs and outputs and that each input combination maps to a unique output combination.

One possible reversible full adder circuit is the Toffoli gate-based design:

```
            a  ─────────→│       ├─ a
                         │       │
            b  ────■────→│       ├─ b
                         │       │
          c_in  ────■───→│       ├─ sum
                         └───┬───┘
                             │
                             │
                    └────────■─ c_out
```

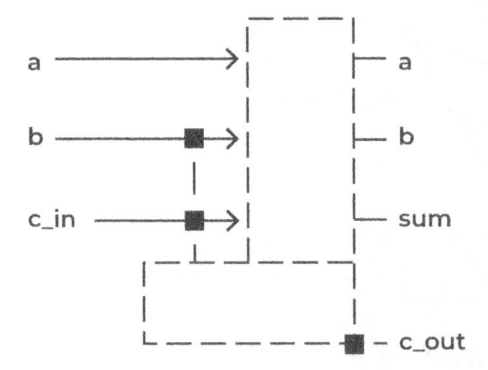

In this circuit, the Toffoli gate (shown as a box with three inputs and outputs) is used to implement the reversible full adder. The inputs are the two operands (a and b) and the carry-in (c_in), while the outputs are the original operands (a and b), the sum, and the carry-out (c_out).

Example: Reversible Parity Checker

A parity checker is a circuit that determines whether the number of 1's in a binary string is even or odd. To create a reversible parity checker, we can use a cascade of controlled-NOT (CNOT) gates:

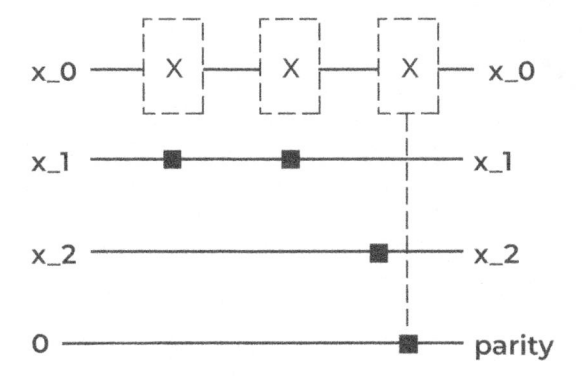

```
 x_0 ──┤ X ├──┤ X ├──┤ X ├── x_0

 x_1 ────■──────■───────────── x_1

 x_2 ──────────────────■────── x_2

  0  ──────────────────────■── parity
```

In this circuit, the CNOT gates are used to compute the parity of the input bits (x_0, x_1, and x_2). The parity bit is initially set to 0, and it is flipped whenever an input bit is 1. The final state of the parity bit represents the parity of the input string (0 for even parity, 1 for odd parity). The original input bits are preserved, ensuring the reversibility of the circuit.

Reversible logic synthesis is an active area of research in quantum computing, with various techniques and algorithms being developed to create efficient and optimized reversible circuits. Some common approaches include gate-based synthesis, template matching, and transformation-based synthesis.

MAPPING TO AVAILABLE GATES:

Mapping to available gates is the process of transforming a quantum circuit to use only the gates supported by the target quantum hardware. This is necessary because different quantum platforms may have different native gate sets, and not all gates can be directly implemented on every hardware. The goal of mapping is to find an equivalent circuit that uses only the available gates while minimizing the number of additional gates and qubits required.

Example: Mapping SWAP Gate

The SWAP gate is a two-qubit gate that exchanges the states of two qubits. However, some quantum hardware may not directly support the SWAP gate. In such cases, we need to map the SWAP gate to a sequence of available gates.

One common decomposition of the SWAP gate uses three controlled-NOT (CNOT) gates:

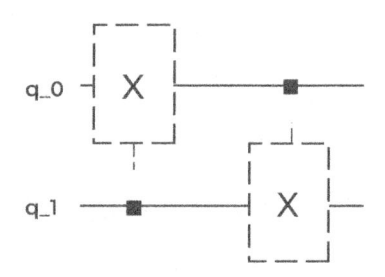

In this decomposition, the first CNOT gate (with q_0 as the control and q_1 as the target) copies the state of q_0 to q_1. The second CNOT gate (with q_1 as the control and q_0 as the target) copies the original state of q_1 to q_0, while also modifying the state of q_1. The third CNOT gate (with q_0 as the control and q_1 as the target) restores the state of q_1 to its original value.

Example: Mapping Toffoli Gate

The Toffoli gate (CCNOT) is a three-qubit gate that performs a controlled-controlled-NOT operation. However, many quantum hardware platforms do not directly support the Toffoli gate. In such cases, we need to map the Toffoli gate to a sequence of available gates.

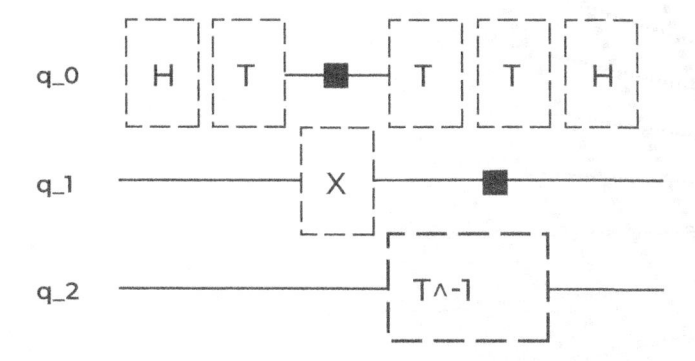

```
q_0 ──[H]──[T]──■──[T]──[T]──[H]──
                │
q_1 ────────────[X]──────■────────
                         │
q_2 ─────────────────[T^-1]───────
```

One possible decomposition of the Toffoli gate uses six controlled-NOT (CNOT) gates and nine single-qubit rotation gates:

In this decomposition, the Toffoli gate is replaced by a sequence of CNOT gates and single-qubit rotation gates (H, T, and T^-1). This decomposition allows the Toffoli gate to be implemented using only the gates available on the target hardware.

Mapping to available gates is a crucial step in the compilation of quantum circuits for execution on real quantum hardware. The choice of mapping depends on the specific hardware constraints, such as the native gate set, qubit connectivity, and error rates. Efficient mapping techniques aim to minimize the overhead in terms of additional gates and qubits while preserving the functionality of the original circuit.

ANCILLA QUBIT REDUCTION

Ancilla qubits, also known as auxiliary qubits, are additional qubits used in quantum circuits to facilitate certain operations or computations. They are not part of the input or output of the algorithm but are used to store intermediate results or perform temporary computations. While ancilla qubits can be useful for implementing complex operations, they also increase the size of the quantum circuit and may introduce additional errors. Therefore, reducing the number of ancilla qubits is an important optimization technique in quantum circuit design.

Example: Reducing Ancilla Qubits in Quantum Adders

Quantum adders are circuits that perform addition on quantum registers. A straightforward implementation of a quantum adder may require several ancilla qubits to store carry bits and perform intermediate computations. However, by using reversible logic gates and carefully designing the circuit, it is possible to reduce the number of ancilla qubits needed.

Consider a simple quantum adder that adds two 2-qubit registers:

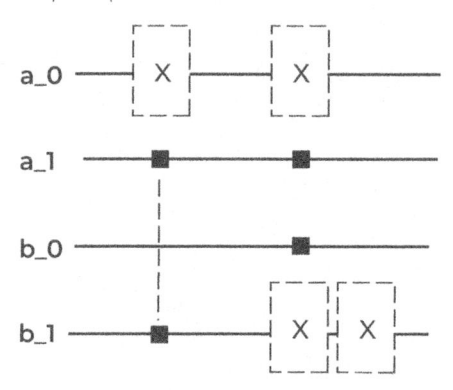

```
a_0 ──┤ X ├──┤ X ├──────
       └─┘   └─┘
a_1 ────■─────■─────────
        │     │
b_0 ────┼─────■─────────
        │
b_1 ────■────┤ X ├┤ X ├─
             └─┘ └─┘
```

In this circuit, the two input registers (a and b) are added using a series of CNOT gates, and the result is stored in the b register. However, this circuit requires an additional ancilla qubit to store the carry bit.

By using the Toffoli gate (CCNOT) and rearranging the gates, we can eliminate the need for the ancilla qubit:

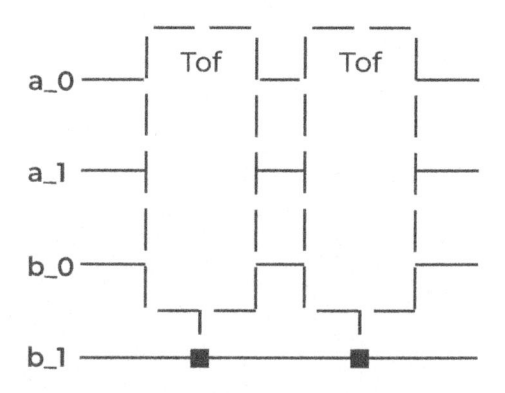

```
a_0 ──┤ Tof ├──┤ Tof ├──
a_1 ──┤     ├──┤     ├──
b_0 ──┤     ├──┤     ├──
b_1 ─────■────────■─────
```

In this optimized circuit, the Toffoli gates are used to compute the carry bits directly, without the need for an ancilla qubit. This reduces the total number of qubits required for the adder circuit.

AUTOMATED OPTIMIZATION:

Powerful optimizers within quantum software packages automate gate count and depth reduction, simplifying optimization and complementing manual design efforts. Heuristic algorithms, such as evolutionary computing and reinforcement learning, effectively navigate large optimization spaces. Optimization frameworks that integrate across software stacks contribute to improving circuits globally.

Automated optimization holds the promise of enabling circuit designs that may be infeasible manually. Similar to classical EDA, the capability for automated optimization is likely to become mandatory for synthesizing complex quantum programs. However, its performance heavily relies on high-quality cost metrics and constraint guidance from designers.

FIDELITY CONSIDERATIONS:

Fidelity is a measure of the accuracy or quality of a quantum operation or circuit. It quantifies how closely the actual output of a quantum system matches the expected or ideal output. Fidelity is an important consideration in quantum circuit optimization, as it directly impacts the reliability and usefulness of the quantum computation.

There are several factors that can affect the fidelity of a quantum circuit:

1. Gate errors: Quantum gates are not perfect and may introduce errors due to imperfect control, calibration, or decoherence. The fidelity of a quantum gate is a measure of how closely it approximates the ideal unitary operation.

2. Measurement errors: Qubit measurements are also subject to errors, such as readout errors or state preparation errors. The fidelity of a measurement is a measure of how accurately it distinguishes between the basis states (e.g., $|0\rangle$ and $|1\rangle$).

3. Decoherence: Qubits are susceptible to decoherence, which is the loss of quantum information due to interactions with the environment. Decoherence can cause the quantum state to evolve in unintended ways, reducing the fidelity of the computation.

4. Crosstalk: In multi-qubit systems, crosstalk can occur when the operation of one qubit inadvertently affects the state of neighboring qubits. Crosstalk can introduce errors and reduce the fidelity of the circuit.

5. Circuit depth: The depth of a quantum circuit, which is the number of time steps required to execute the circuit, can also impact fidelity. Deeper circuits are more susceptible to errors due to decoherence and gate imperfections.

When optimizing quantum circuits, it is important to consider fidelity as a key performance metric. The goal is to find a circuit that not only minimizes the number of gates or depth but also maximizes the fidelity of the computation.

Example: Fidelity-aware Circuit Optimization

Consider a quantum circuit that implements a simple quantum algorithm, such as the Deutsch-Jozsa algorithm. The circuit consists of a series of Hadamard gates, an oracle function, and measurements:

To optimize this circuit for fidelity, we can consider several strategies:

1. Gate optimization: We can replace the generic Hadamard gates with optimized versions that have higher fidelity, such as

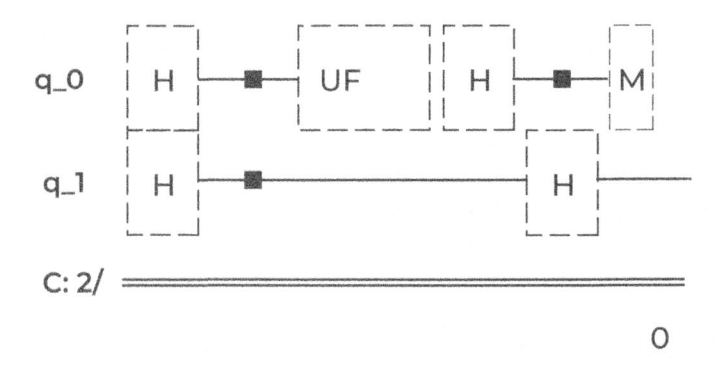

```
q_0  ─H─■──UF──H─■──M
                         
q_1  ─H─■──────────H─
                         
C: 2/ ══════════════════
                      0
```

dynamically corrected gates or composite pulse sequences.

2. Oracle optimization: The oracle function can be optimized to minimize the number of gates and depth, reducing the opportunities for errors to accumulate. This may involve using ancilla qubits, reversible logic synthesis, or automated circuit optimization techniques.

3. Measurement optimization: We can use readout error mitigation techniques, such as readout calibration or post-processing, to improve the fidelity of the measurements.
4. Layout optimization: If the circuit is mapped to a specific quantum hardware architecture, we can optimize the layout of the qubits to minimize crosstalk and maximize the fidelity of the two-qubit gates.
5. Error correction: We can incorporate quantum error correction codes, such as the surface code or the color code, to detect and correct errors during the execution of the circuit. This can significantly improve the overall fidelity of the computation, at the cost of additional qubits and gates.

OPTIMIZATION CHALLENGES:

Finding globally optimal solutions quickly becomes intractable as circuit size increases. Heuristics, approximations, and design constraints are employed to address optimization challenges.

Numerous equivalent circuit form rearrangements exist exponentially, and noise, along with model inaccuracies, complicates assessment. Constraints narrow down options but at the cost of limiting potential. Tradeoffs always exist between optimality guarantees, scalability, and design effort.

Practical quantum circuit optimization requires navigating tensions between solution quality, runtime, and designer effort. Insights from theoretical computer science offer optimization techniques that align with real-world engineering challenges. Similar to classical heuristics, developing robust quantum optimizers relies heavily on intuition derived from experience.

OBJECTIVES PRIORITIZATION

In navigating multidimensional optimization spaces, it becomes imperative to prioritize certain circuit objectives over others. This prioritization is driven by preferences based on the application and hardware, guiding the emphasis on the optimization process.

For instance, in the realm of quantum simulation, the focus leans towards minimizing qubit resources rather than reducing depth. On the other hand, error-corrected algorithms prioritize the maximization of fidelity over minimizing gate counts. To strike a balance, optimizers integrate weighted guidance, enabling them to navigate tradeoffs judiciously.

Defining optimization priorities is crucial for directing efforts efficiently, especially when faced with limited resources. It's essential to acknowledge that not all objectives can be simultaneously optimized. Therefore, providence guidance through weighting, sequencing, and constraints becomes paramount, allowing for the identification of Pareto-optimal approximations.

OPTIMIZATION LIMITS

Fundamental physical constraints bound the performance achievable through gate optimization. Limits include uncertainty principles and bounds on parallelism.

The time-energy uncertainty principle lower-bounds the evolution time of a desired unitary operation. Topology and data movement bound parallel execution. Approaching these limits requires co-design of hardware, gates, and error correction.

Practical quantum gate optimization must operate within known physics restrictions. While daunting, these limits imply remaining headroom to improve as technology matures. Co-optimization and deriving empirical performance bounds for particular hardware will help guide expectations and inform potential.

HARDWARE CO-DESIGN

Optimizing quantum gate sets should be tailored to account for target hardware capabilities and limitations. This approach facilitates efficient execution on actual devices.

Considerations for device physics and fabrication constraints are crucial in determining viable native operations. Gate co-design, specifically tailored for the characteristics of specific qubit modalities and interconnects, enhances mapping reliability. Accounting for calibration accuracy and hardware drift further boosts robustness.

Quantum computer engineering necessitates the co-design of algorithms, software, circuits, and hardware collectively rather than in isolation. Holistic co-optimization distributes tradeoffs across the stack to maximize total performance. The selection of the right gates for algorithms depends intrinsically on what hardware can reliably provide.

PARAMETERIZED CIRCUITS

Parametric quantum circuits have tunable gate settings enabling optimization over families of related implementations.

Examples include variational quantum eigensolvers and quantum neural networks. Parameterized circuits are classically optimized toward goals like energy minimization and model accuracy by tuning gate variables. Parameterized circuits provide powerful optimization abstractions by encoding solution spaces compactly through gate settings. This facilitates exploring optimizations programmatically through gradient descent or evolutionary algorithms to uncover high-quality circuit instances from exponentially large spaces.

Quantum Circuit Simulation and Verification

Simulating quantum circuit behaviour is essential for design verification and performance evaluation. Simulation facilitates debugging, refinement, and validation before costly quantum hardware execution.

QUANTUM STATE REPRESENTATION

Qubit state vectors, density matrices, and stabilizer formalism enable simulating quantum state evolution under gate operations. Different representations have trade-offs. Qubit state vectors directly model superposition and entanglement. Density matrices also capture mixed states and incoherent noise processes. Stabilizer methods focus on Clifford gate circuits. Representations carry different simulatability on classical computers. Choosing an appropriate state representation is critical for efficient simulation. Vectors scale poorly with qubits but readily capture coherence. Density matrices model ensemble noise yet square simulation costs. Stabilizers efficiently simulate limited circuits. Hybrid approaches combine strengths across formalisms.

STATE PROPAGATION

Applying gate unitaries sequentially to initial states propagates the quantum state through the circuit model. This numerically simulates circuit dynamics. Basic simulations multiply successive gate matrices with state vectors or density operators. Optimized implementations apply gates symbolically, propagating states without explicit matrices. Propagation tracks quantum evolution. Iterative state propagation forms the computational core of quantum circuit simulation. Performant propagation relies on data structures and algorithms reflecting matrix sparsity and circuit connectivity structure. Numerical stability considerations also arise during lengthy simulations.

PERFORMANCE EVALUATION

Simulators calculate figures of merit, including state probabilities, entanglement, purity, and fidelity. This evaluates circuit performance before hardware implementation. Metrics quantify aspects like output accuracy, noise resilience, resource costs, and algorithmic convergence. Probabilistic nature necessitates statistical approaches via sampling or ensemble methods. Metrics guide debugging and refinement. Well-defined performance metrics enable standardized evaluation of simulated circuit quality. Incorporating hardware imperfections into idealized simulations provides more realistic assessments. Holistic analysis balancing factors like fidelity, depth, and qubit overhead identifies optimization opportunities.

DEBUGGING

Injecting faults into simulations provides debugging insight. Simulated outputs can be compared with theoretical expectations to identify and correct issues. Controlled fault injection mimics the noise and errors afflicting real hardware. Observing simulation perturbations highlights circuit vulnerabilities. Systematically locating and patching faulty behaviour improves robustness.

Thorough fault testing via simulation facilitates debugging and verification before expensive hardware experiments. Analogous to classical testbenches, constructing representative fault scenarios grants insight into failure modes and guides refinement. Debugging will grow more indispensable as circuit complexity increases.

NOISE INJECTION

Real hardware imperfections can be modelled by injecting noise into simulations. This provides more realistic estimates of practical circuit operation.

Common noise models include depolarizing, amplitude damping, phase damping, and dephasing. Statistical noise injection approximates observed device noise. Noisy simulation better predicts true hardware outputs. While adding computational overhead, injecting calibrated hardware noise profiles into simulations offers critical real-world performance estimates. Combining noise with fault injection provides robust circuit validation before hardware implementation. Noisy simulation guides design refinements targeting performance bottlenecks.

CLASSICAL VS. QUANTUM

Classical simulation scales poorly with qubit count due to exponential state space growth. Quantum hardware offers exponential advantages for large-scale circuit simulation.

Classical approximation methods like tensor networks partially alleviate state space explosions. Regardless, current classical machines remain incapable of fully simulating deep quantum circuits over many qubits. Quantum hardware exponentially expands accessible simulation complexity versus classical limitation. However, engineering challenges persist in building high-fidelity quantum platforms that can reliably simulate themselves or other quantum systems. This remains an active research frontier.

APPROXIMATION TECHNIQUES

Methods like tensor networks and branch tracing approximate large circuit simulations using limited classical resources. These trade-offs are accuracy for computational efficiency. Tensor networks exploit matrix product structure to compactly represent states. Branch tracing prioritizes the evolution of significant basis states. There are many simulation approximation strategies.

Practical quantum circuit simulation leverages a spectrum of approximation techniques balancing precision, performance, and complexity. As with classical equivalents, utilizing domain-specific heuristics to focus computation where it matters most extends the reachable simulation scale.

SIMULATOR VALIDATION

Trusted high-quality simulators are vital for correct circuit evaluation. Rigorous validation using analytical and empirical techniques inspires simulator confidence.

Mathematical proofs demonstrate simulator correctness on representative circuits. Cross-checking against other simulators and real hardware provides empirical validation. Unit testing catches regressions during software changes. Building confidence in simulation accuracy is critical, given the central role of simulators in quantum programming today. Validation combines mathematical formalism, scientific principles, software engineering, and metrological determination of simulator uncertainties relative to ground truth hardware behaviour.

SIMULATION LIBRARIES

Common simulation tasks are packaged into reusable open-source libraries integrated with quantum programming frameworks. These aid in efficient simulation development.

Libraries contain optimized components for state representation, propagation, noise models, metrics calculation, decompositions, and approximation techniques. High-performance linear algebra libraries also accelerate simulation.

Robust open-source simulation libraries accelerate development and uptake. Integration with popular programming platforms like Qiskit improves accessibility. As with software engineering, generally, thoughtful library design promotes abstraction, interoperability, and extensibility as quantum computing matures.

HYBRID SIMULATION

Combining classical simulators with real quantum co-processors can boost simulation scale and practical accuracy for complex circuits and algorithms.

Hybrid simulation leverages available qubits to explore portions of the state space that are classically intractable. Feedback between quantum and classical domains combines strengths from each. Hybrid simulation offers a pragmatic engineering solution to expanding quantum simulation capacity. Moving toward fully quantum simulators will require progress on qubit count, connectivity, and error rates. For now, hybrid balancing ameliorates classical vs. quantum resource limitations.

MONTE CARLO METHODS

Randomized sampling provides efficient simulation estimates of qubit measurement probabilities after state propagation.

Many probabilistic runs build outcome histograms, approximating quantum measurement distributions. Convergence to true probabilities occurs asymptotically as the sample count increases. Importance sampling focuses on significant basis states. Monte Carlo approaches affordably approximate quantum state information via statistics gathering. As with classical usages, importance sampling offers variance reduction to accelerate convergence toward accurate results. Enabling cheap estimation of probabilities may drive early applications.

MULTISCALE METHODS

Combining quantum simulators and classical calculations models interacting phenomena operating on vastly different spatial and temporal scales. Timestep splitting apportions fast quantum and slower classical scale updates across synchronization boundary layers. This facilitates modelling the joint evolution of coupled systems at disparate rates.

Hybrid multiscale modelling brings together strengths from quantum and classical techniques best suited to phenomena at their native scales. Chemistry simulation combining electronic quantum effects with molecular dynamics offers an application example of multiscale methods.

BATCH SIMULATION

Grouping many related circuit simulations into batched jobs improves throughput on classical and quantum hardware.

Batching amortizes fixed overheads like compilation across simulations. Vectorized execution and shared intermediates also enhance performance. Parameter sweeps explore wide optimization spaces.

Intelligent batching strategies will unlock efficient simulation at scale. Batch optimization considers tradeoffs between coherence, overhead, and locality. Hybrid simulations may pipeline classically batched preparation, execution, and postprocessing around quantum evaluation.

SIMULATION PERFORMANCE

High-performance simulation utilizes optimized data structures, kernel fusion, parallelism, and hardware acceleration to maximize computational throughput.

Efficient state data layouts reflect sparsity and access patterns. Kernel fusion reduces intermediate allocation. Parallelism leverages SIMD instructions and distributed hardware. GPUs and FPGAs also accelerate.

Practical quantum circuit simulation demands high-performance engineering to address exponential complexity scaling. Hybrid processors combining scalar, vector, spatial, and temporal parallelism provide necessary throughput capabilities as system sizes grow.

CONTINUOUS SIMULATION

Specialized techniques enable the direct simulation of the continuous-time dynamics of always-on quantum systems without circuit discretization.

Continuous simulation is required for analogue quantum computers and simulating continuous dynamical processes natively, avoiding time discretization error. Solving the time-dependent Schrödinger equation propagates continuous evolution.

Support for continuous-time simulation better captures the inherent dynamics of quantum systems versus overly discretizing into time slices. This increases model fidelity for quantum processes involving always-on interactions and controllable drives unsuited to pulsed gate abstraction.

CHAPTER 6
QUANTUM ALGORITHMS IN PRACTICE

Variational Quantum Algorithms for Machine Learning and Optimization

Variational quantum algorithms are a leading approach for near-term quantum applications. These hybrid quantum-classical algorithms work by using a quantum processor to evaluate the cost function for an optimization problem or machine learning model, while a classical optimizer adjusts the parameters to minimize the cost function. The "variational" aspect refers to the classical optimization loop. Key advantages of these algorithms are that they can be run on noisy intermediate-scale quantum (NISQ) devices, do not require full error correction, and provide a pathway to quantum advantage. Variational algorithms are actively being researched and tested for machine learning as well as combinatorial and quantum simulation optimization.

QUANTUM NEURAL NETWORKS

Quantum neural networks (QNNs) are a type of variational quantum algorithm that combines the principles of artificial neural networks with the capabilities of quantum computing. QNNs aim to harness the power of quantum superposition and entanglement to learn complex patterns and representations from data.

The basic structure of a QNN consists of a layered network of quantum gates, similar to the layers in a classical neural network. Each layer applies a series of parameterized quantum gates to the input qubits, transforming their state according to the learned parameters. The output of the QNN is typically a measurement on one or more qubits, which can be interpreted as a classification label or a continuous value.

Example: QNN for Binary Classification

Consider a binary classification task, where the goal is to classify input data into two classes (e.g., 0 or 1). A QNN for this task can be constructed as follows:

1. Encode the input data: The input features are encoded into the state of the qubits using a suitable encoding scheme, such as amplitude encoding or angle encoding.

2. Apply parameterized quantum layers: A series of parameterized quantum gates, such as rotations and entangling gates, are applied to the qubits. The parameters of these gates are learned during the training process.

3. Measure the output qubit: The state of a designated output qubit is measured, and the measurement outcome is interpreted as the predicted class label (e.g., 0 or 1).

4. Train the QNN: The QNN is trained using a hybrid quantum-classical optimization algorithm, such as the variational quantum eigensolver (VQE) or the quantum approximate optimization algorithm (QAOA). The parameters of the quantum gates are updated iteratively to minimize a loss function, such as the binary cross-entropy.

Here's a simplified example of a QNN circuit for binary classification:

```
q_0 ─── H ── RZ(0_1) ──■──■────────── H ── M
                       │  │
q_1 ─── H ── RZ(0_2) ──X──X────── M
                                  │
C: 2/ ════════════════════════════╧═══
                                  0
```

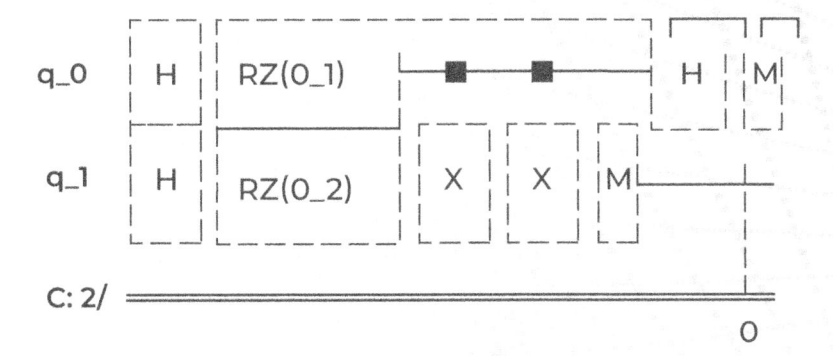

In this example, the input features are encoded using Hadamard gates (H), and the parameterized quantum layer consists of rotation gates (RZ) and controlled-X gates (CX). The output qubit (q_0) is measured, and the measurement outcome is used to predict the class label.

QUANTUM ALGORITHMS FOR LINEAR ALGEBRA

Linear algebra is a fundamental tool in many areas of science and engineering, including machine learning, optimization, and numerical simulations. Quantum computers have the potential to perform certain linear algebra tasks more efficiently than classical computers, thanks to their ability to manipulate vast quantum state spaces.

Several quantum algorithms have been developed for linear algebra tasks, such as solving systems of linear equations, eigenvalue estimation, and matrix inversion. These algorithms typically rely on the quantum linear systems algorithm (QLSA) or its variants as a key subroutine.

Example: Quantum Linear Systems Algorithm (QLSA)

The QLSA, also known as the HHL algorithm (named after its inventors Harrow, Hassidim, and Lloyd), is a quantum algorithm for solving systems of linear equations of the form $Ax = b$, where A is a Hermitian matrix and b is a vector.

The QLSA consists of three main steps:

1. Quantum state preparation: The vector b is encoded into a quantum state |b⟩ using techniques such as amplitude encoding or qRAM (quantum random access memory).

2. Quantum phase estimation: The eigenvalues and eigenvectors

of the matrix A are estimated using the quantum phase estimation algorithm. This step requires the ability to efficiently simulate the matrix exponential e^(iAt).

3. Quantum state tomography: The solution vector x is extracted from the quantum state |x⟩ using techniques such as quantum state tomography or amplitude amplification.

Here's a simplified example of the QLSA circuit:

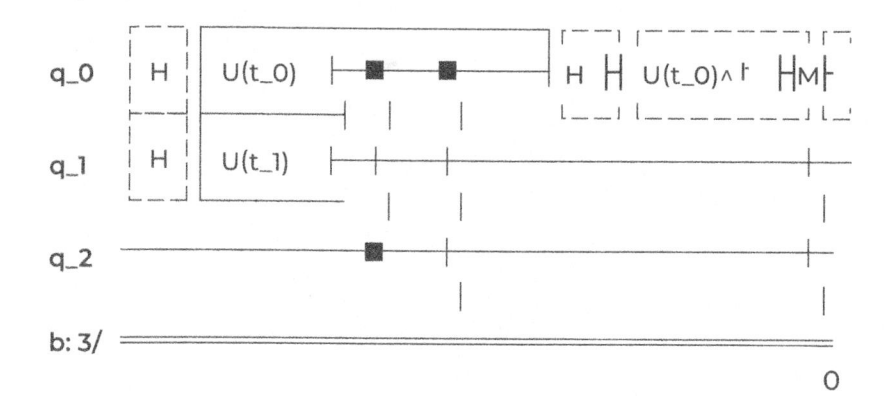

```
q_0   | H || U(t_0) |----■----■---------| H || H || U(t_0)^t || H ||M|-
      |___||_____|    |    |         |___||___||_____||___||_|
q_1   | H || U(t_1) |----|----|---------------------------------------
      |___||_____|    |    |
                         |    |
q_2   -------------------■----|---------------------------------------
                              |
b: 3/ ================================================================
                                                                     0
```

In this example, the matrix A is encoded into the controlled-U gates, where U(t) = e^(iAt). The eigenvectors of A are encoded into the quantum state |b⟩, which is prepared using amplitude encoding. The quantum phase estimation algorithm is applied to estimate the eigenvalues, and the solution vector |x⟩ is extracted using quantum state tomography.

QUANTUM COMBINATORIAL OPTIMIZATION

Quantum combinatorial optimization is a field that explores the use of quantum algorithms to solve complex optimization problems, such as the traveling salesman problem, the maximum cut problem, and the graph coloring problem. These problems are known to be NP-hard, meaning that they are computationally intractable for classical computers as the problem size grows.

Quantum algorithms, such as the quantum approximate optimization algorithm (QAOA) and the quantum adiabatic algorithm (QAA), have shown promise in providing a speedup over classical algorithms for certain instances of these problems.

Example: Quantum Approximate Optimization Algorithm (QAOA)

The QAOA is a hybrid quantum-classical algorithm that combines the principles of quantum annealing and classical optimization to solve combinatorial optimization problems.

The QAOA consists of two main steps:

1. Quantum circuit construction: A parameterized quantum circuit is constructed based on the problem instance. The circuit typically consists of alternating layers of problem-specific gates (e.g., the phase separator) and mixing gates (e.g., the transverse field).

2. Classical optimization: The parameters of the quantum circuit are optimized using a classical optimization algorithm, such as gradient descent or Bayesian optimization, to maximize the expected value of the problem Hamiltonian.

Here's a simplified example of a QAOA circuit for the maximum cut problem:

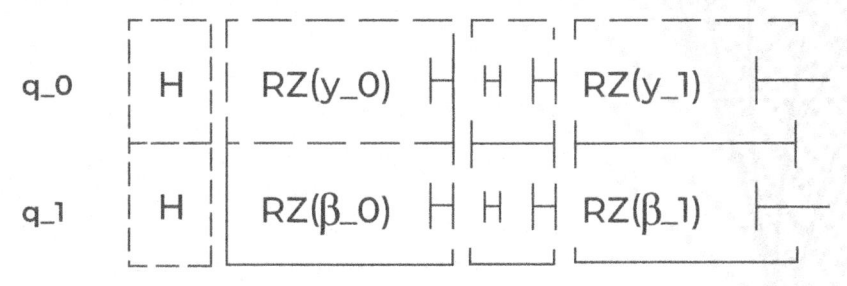

```
q_0 ─┤ H ├─┤ RZ(y_0) ├─┤ H ├─┤ RZ(y_1) ├─
q_1 ─┤ H ├─┤ RZ(β_0) ├─┤ H ├─┤ RZ(β_1) ├─
```

In this example, the problem-specific gates are the RZ gates with parameters y_i, which encode the graph structure and edge weights. The mixing gates are the RZ gates with parameters $β\_i$, which introduce quantum fluctuations and enable the exploration of the solution space.

The QAOA has been shown to provide a quadratic speedup over classical algorithms for certain instances of the maximum cut problem and has been successfully demonstrated on small-scale quantum hardware.

IMPLEMENTATION AND OUTLOOK

Leading quantum computing companies like D-Wave, Rigetti, and IonQ are actively working on implementing variational quantum algorithms on their hardware platforms. Partnerships with classical computing companies are also being pursued to enable hybrid deployments.

While current implementations are still small in scale, rapid iterative improvements to quantum processors, compilers, and classical co-processor integration are being made. If theoretical speedups can be realized in practice, variational quantum algorithms could be among the first to achieve true quantum advantage for impactful real-world applications. However, algorithmic improvements must also continue to enhance scalability, noise resilience, and solution quality.

Overall, variational quantum algorithms represent a promising bridge to running complex quantum applications on near-term quantum devices. Continued hardware and software co-design efforts will help mature these algorithms from proofs-of-concept to practical tools for optimization and machine learning over the coming years.

QUANTUM CIRCUIT MODEL

The quantum circuit model is the predominant framework for designing variational quantum algorithms. In this model, quantum computations are broken down into sequences of fundamental quantum logic gates. The gates are applied to qubits prepared in initial states to perform state evolution and transformations. Measurement at the end of the circuit generates the output. Parameterized circuits have variable coefficients within certain gates that can be tuned to optimize algorithm performance.

Quantum circuit design is key to realizing effective variational algorithms. Architectures must balance expressiveness, trainability, and hardware efficiency. Common design patterns include layered networks of parameterized gates to enact quantum feature maps. Compiling circuits to target hardware while minimizing noise and qubit connectivity constraints is also essential for practical implementations.

HYBRID QUANTUM-CLASSICAL OPTIMIZATION

Variational quantum algorithms utilize a hybrid optimization process between quantum and classical hardware. The quantum processor executes the quantum circuit to evaluate objective functions. Then, classical optimizers such as gradient descent use this information to tune the quantum circuit parameters for subsequent iterations.

The aim is to leverage the strengths of both computing paradigms. The quantum circuit provides an efficient evaluation of costs for optimization problems or machine learning models. Meanwhile, the classical processor handles outer loops for iterative optimization and final results processing.

Tight integration and interleaving between quantum and classical hardware can enable fast, low-latency hybrid computations. As variational algorithms scale, efficient hybrid execution will become increasingly crucial.

TRAINING CHALLENGES

A key challenge with variational quantum algorithms is training the quantum circuits. Issues such as barren plateaus, local optima, and stochastic gradients can hamper learning. Noise and decoherence also impair training.

Strategies to aid training include smarter classical optimizers, gradient estimation techniques like parameter shift rules, and circuit architecture improvements. Robust training procedures will be critical to unlocking the potential of variational quantum algorithms.

DATA ENCODING

Data encoding is the process of converting classical data into a format that can be processed by a quantum computer. This is a crucial step in many quantum machine learning algorithms, as it allows classical data to be represented and manipulated in a quantum state space.

There are several methods for encoding data into quantum states, each with its own advantages and limitations. Some common encoding methods include:

Amplitude encoding

In amplitude encoding, the data is encoded into the amplitudes of the quantum state vector. For example, a d-dimensional classical vector x can be encoded into a log(d)-qubit quantum state |x⟩

as follows: |x⟩ = Σ_i x_i |i⟩ / ||x|| where |i⟩ are the computational basis states, and ||x|| is the Euclidean norm of x. Amplitude encoding allows for exponentially compact representations of classical data, as a d-dimensional vector can be encoded into log(d) qubits. However, preparing arbitrary amplitude-encoded states can be challenging, requiring techniques such as quantum random access memory (qRAM) or quantum state preparation circuits.

Basis encoding

In basis encoding, the data is encoded into the computational basis states of the qubits. For example, a binary string x of length n can be encoded into an n-qubit quantum state |x⟩ as follows: |x⟩ = |x_1⟩ ⊗ |x_2⟩ ⊗ ... ⊗ |x_n⟩ where |x_i⟩ is the computational basis state corresponding to the i-th bit of x, and ⊗ denotes the tensor product. Basis encoding is straightforward to implement, as it only requires single-qubit gates to prepare the desired state. However, it requires a number of qubits equal to the size of the classical data, which can limit its scalability.

Angle encoding

In angle encoding, the data is encoded into the rotation angles of single-qubit gates. For example, a real-valued vector x of length n can be encoded into an n-qubit quantum state |x⟩ as follows: |x⟩ = ∏_i Ry(x_i) |0⟩ where Ry(x_i) is a rotation gate around the y-axis by an angle proportional to x_i. Angle encoding allows for a compact representation of real-valued data, as each value is encoded into a single rotation gate. However, it may require a large number of gates to prepare the desired state, especially for high-dimensional data.

Here's an example of encoding a 4-dimensional binary vector using basis encoding:

from qiskit import QuantumCircuit

# Classical data
x = [1, 0, 1, 1]

# Create a quantum circuit with 4 qubits
qc = QuantumCircuit(4)

```
# Encode the classical data into the quantum state
for i in range(4):
    if x[i] == 1:
        qc.x(i)


# Print the quantum circuit
print(qc)
Output:
```

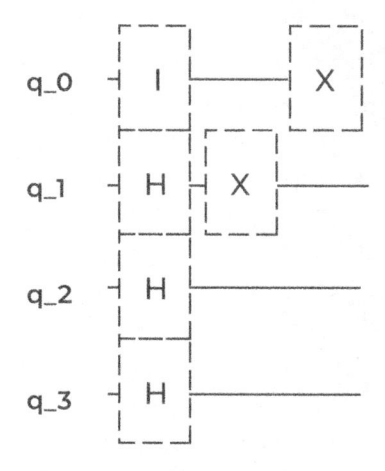

```
q_0  ─ I ──────── X ─
                      
q_1  ─ H ── X ───────
                      
q_2  ─ H ────────────
                      
q_3  ─ H ────────────
```

In this example, the classical binary vector [1, 0, 1, 1] is encoded into the computational basis states of 4 qubits using single-qubit X gates. The resulting quantum state is |1011⟩.

The choice of encoding method depends on the specific requirements of the quantum algorithm, such as the type and size of the classical data, the available quantum resources, and the desired computational complexity.

As quantum machine learning continues to advance, new encoding methods are being developed to address the limitations of existing techniques and enable the efficient processing of large-scale, high-dimensional data on quantum computers.

MACHINE LEARNING APPLICATIONS

Quantum machine learning is an emerging field that explores the use of quantum algorithms to solve machine learning problems. By leveraging the unique properties of quantum computation, such as superposition, entanglement, and interference, quantum machine learning algorithms have the potential to provide speedups and improvements over classical machine learning algorithms.

Some potential applications of quantum machine learning include:

1. Classification: Quantum algorithms can be used to train and evaluate classifiers, such as support vector machines and neural networks, on classical or quantum data. For example, the quantum support vector machine (QSVM) algorithm uses a quantum kernel function to map the input data into a high-dimensional feature space, enabling the efficient classification of complex datasets.

2. Clustering: Quantum algorithms can be used to partition data into groups based on similarity or distance metrics. For example, the quantum k-means algorithm uses a quantum subroutine to estimate the distances between data points and cluster centroids, enabling the efficient clustering of large datasets.

3. Dimensionality reduction: Quantum algorithms can be used to extract the most informative features from high-dimensional data, reducing the computational complexity of downstream machine learning tasks. For example, the quantum principal component analysis (qPCA) algorithm uses a quantum subroutine to estimate the eigenvalues and eigenvectors of the covariance matrix, enabling the efficient compression and visualization of large datasets.

4. Generative models: Quantum algorithms can be used to learn the underlying probability distribution of a dataset and generate new samples from that distribution. For example, the quantum Boltzmann machine (QBM) algorithm uses a quantum circuit to model the joint probability distribution of visible and hidden variables, enabling the efficient generation of realistic samples.

5. Reinforcement learning: Quantum algorithms can be used to learn optimal policies for decision-making in complex environments. For example, the quantum Q-learning algorithm uses a quantum subroutine to estimate the action-value function, enabling the efficient exploration and exploitation of large state spaces.

Here's an example of using the quantum support vector machine (QSVM) algorithm for binary classification:

from qiskit import Aer, QuantumCircuit

```python
from qiskit.aqua import QuantumInstance
from qiskit.aqua.algorithms import QSVM
from qiskit.aqua.components.feature_maps import SecondOrderExpansion

# Load and preprocess the data
X_train, y_train, X_test, y_test = load_data()

# Define the feature map and quantum instance
feature_map = SecondOrderExpansion(num_qubits=2, depth=2)
backend = Aer.get_backend('qasm_simulator')
quantum_instance = QuantumInstance(backend, shots=1024)

# Define the QSVM model
qsvm = QSVM(feature_map, quantum_instance)

# Train the QSVM model
qsvm.fit(X_train, y_train)

# Evaluate the QSVM model
accuracy = qsvm.score(X_test, y_test)
print(f'Accuracy: {accuracy:.2f}')
```

In this example, the QSVM algorithm is used to train a binary classifier on a given dataset. The input data is first mapped into a higher-dimensional feature space using a second-order expansion feature map. The QSVM model is then trained on the quantum feature map using a quantum circuit and a classical optimizer. Finally, the trained model is evaluated on a test set to measure its accuracy.

While still in its early stages, quantum machine learning has shown promising results on a variety of tasks, from image recognition and natural language processing to drug discovery and financial modeling. As quantum hardware and software continue to improve, the potential of quantum machine learning to solve real-world problems is expected to grow, leading to new applications and insights in fields such as healthcare, finance, and artificial intelligence.

COMBINATORIAL OPTIMIZATION

Combinatorial optimization is a branch of optimization that deals with problems where the solution space is discrete and often very large. Examples of combinatorial optimization problems include the traveling salesman problem, the maximum cut problem, and the graph coloring problem.

Quantum algorithms have shown promise in solving certain types of combinatorial optimization problems more efficiently than classical algorithms. Some popular quantum algorithms for combinatorial optimization include:

Quantum Approximate Optimization Algorithm (QAOA)

QAOA is a hybrid quantum-classical algorithm that combines a parameterized quantum circuit with a classical optimization routine to find approximate solutions to combinatorial optimization problems. The quantum circuit is used to explore the solution space and generate candidate solutions, while the classical optimizer is used to adjust the parameters of the quantum circuit based on the quality of the solutions. Example: Maximum Cut Problem The maximum cut problem is a well-known combinatorial optimization problem where the goal is to partition the vertices of a graph into two sets such that the number of edges between the sets is maximized. QAOA can be used to find approximate solutions to the maximum cut problem by encoding the problem instance into a cost Hamiltonian and using a parameterized quantum circuit to minimize the expectation value of the Hamiltonian.

Quantum Annealing

Quantum annealing is a heuristic optimization algorithm that uses a quantum system to explore the solution space of a combinatorial optimization problem. The algorithm works by encoding the problem instance into a quantum Hamiltonian and slowly evolving the system from an initial state to a final state that represents the optimal solution. Quantum annealing has been implemented on specialized quantum hardware, such as D-Wave's quantum annealers. Example: Traveling Salesman Problem The traveling salesman problem is a classic combinatorial optimization problem where the goal is to find the shortest possible route that visits each city exactly once and returns to the starting city. Quantum annealing can be used to find approximate solutions to the traveling salesman problem by encoding the problem instance into a quadratic unconstrained binary optimization (QUBO) problem and using a quantum annealer to minimize the energy of the QUBO.

Variational Quantum Eigensolver (VQE)

VQE is a hybrid quantum-classical algorithm that uses a parameterized quantum circuit to prepare an ansatz state and a classical optimizer to minimize the expectation value of a problem Hamiltonian. While originally developed for quantum chemistry problems, VQE has also been applied to combinatorial optimization problems by encoding the problem instance into a suitable Hamiltonian. Example: Graph Coloring Problem The graph coloring problem is a combinatorial optimization problem where the goal is to assign colors to the vertices of a graph such that no two adjacent vertices have the same color, using the minimum number of colors possible. VQE can be used to find approximate solutions to the graph coloring problem by encoding the problem instance into a Hamiltonian and using a parameterized quantum circuit to minimize the energy of the Hamiltonian.

Here's an example of using QAOA to solve the maximum cut problem:

```
import networkx as nx
from qiskit import Aer, QuantumCircuit
from qiskit.aqua import QuantumInstance
from qiskit.aqua.algorithms import QAOA
from qiskit.aqua.components.optimizers import SPSA

# Create a random graph
G = nx.gnm_random_graph(5, 7)

# Define the QAOA instance
backend = Aer.get_backend('qasm_simulator')
quantum_instance = QuantumInstance(backend, shots=1024)
optimizer = SPSA(maxiter=100)
qaoa = QAOA(optimizer, quantum_instance, p=2)

# Run QAOA on the maximum cut problem
```

result = qaoa.run(G)

# Print the results
print(f'Optimal cut value: {result["optimal_value"]:.2f}')
print(f'Optimal cut: {result["optimal_cut"]}')

In this example, a random graph with 5 vertices and 7 edges is generated using NetworkX. The QAOA algorithm is then instantiated with a quantum simulator backend, a classical optimizer (SPSA), and a depth parameter (p=2). The QAOA algorithm is run on the maximum cut problem instance defined by the graph, and the optimal cut value and vertex partition are printed.

QUANTUM CHEMISTRY SIMULATION

Quantum chemistry seeks to model chemical systems from first principles through physics-based simulation. This is infeasible on classical computers. However, quantum computers can represent precise molecular wavefunctions to enable accurate quantum chemistry simulations.

With sufficient scale, quantum simulations could transform drug design, battery development, catalyst discovery, and more. Hybrid quantum algorithms show promise for delivering advanced quantum chemistry capabilities sooner.

QUANTUM MACHINE LEARNING FOR SCIENCE

In addition to optimization, machine-learning techniques powered by quantum circuits could accelerate scientific discoveries. Possible applications include chemical design, particle physics, cosmology, and more.

Hybrid schemes marrying quantum and classical machine learning algorithms may offer performance benefits over either alone. This emerging niche within quantum computing could greatly expand how algorithms enhance scientific exploration.

HARDWARE PLATFORMS

Leading quantum computing companies like Google, IBM, Rigetti, and IonQ are actively working to implement variational quantum algorithms on their qubit technologies. Partnerships to enable hybrid quantum-classical co-processing are also underway.

While current quantum processors remain limited, rapid iterative improvements in hardware, software, and integration will help mature variational algorithms from proofs-of-concept to practical applications.

SOFTWARE STACKS

To support variational algorithm development, dedicated software stacks are emerging. These provide hybrid programming frameworks, simulators, compilers, optimizers, and other tools for creating end-to-end solutions.

Integrated software environments that tightly link algorithm design, optimization, and hardware execution will help streamline the development and deployment of variational quantum applications.

Quantum Approximate Optimization Algorithm (QAOA) for Combinatorial Problems

Combinatorial optimization involves finding the optimal solution from a finite but extremely large set of possibilities. Problems like scheduling, protein folding, and network routing belong to this class. The quantum approximate optimization algorithm (QAOA) offers a pathway to leverage quantum effects for enhanced combinatorial optimization. In QAOA, a cost Hamiltonian encodes the optimization problem's objective function. By cleverly choosing a mixing Hamiltonian, quantum annealing can help the system tunnel through local minima and converge toward optimal or near-optimal solutions. QAOA is also a hybrid algorithm, using a quantum subroutine with a classical outer loop. Research is advancing QAOA, bringing it closer to mainstream viability for critical industrial applications.

OVERVIEW OF QAOA

The Quantum Approximate Optimization Algorithm (QAOA) stands out as a prominent hybrid quantum-classical algorithm designed for approximating solutions to combinatorial optimization problems. Its operational framework involves encoding the optimization's cost function into a "cost Hamiltonian" and selecting an optimized "mixer Hamiltonian." The quantum processor then navigates the intricate energy landscape of potential solutions, aiming to converge towards optimal solutions.

The quantum processor generates a superposition of states, evolving them based on the combined cost and mixer Hamiltonians. After repeated mixing and evolution, the measurement produces candidate solutions. These solutions are fed into a classical optimizer that refines the mixer parameters, steering towards better solutions in subsequent iterations. Despite its conceptual simplicity, the effectiveness of QAOA hinges on the intricate task of choosing suitable encodings and mixer Hamiltonians tailored to the specific problem and hardware, posing a significant but surmountable challenge. Successful implementation of QAOA can yield high-quality approximate solutions to discrete optimization problems, which are often intractable for classical algorithms.

APPLICATIONS TO COMBINATORIAL PROBLEMS

QAOA finds particular relevance in solving combinatorial problems where determining the global minimum of a complex cost function is challenging. This includes applications in scheduling, protein folding, portfolio optimization, network routing, and more. These problems can be encoded as quadratic unconstrained binary optimizations (QUBOs) or Ising Hamiltonians, subsequently solved through QAOA mixing and measurement.

The adaptability of QAOA across different encodings makes it applicable across various domains such as logistics, finance, engineering, and computational biology. Identifying problems amenable to QAOA's strengths, along with tailoring mixing operators and circuits for maximal performance, constitutes an active area of research. Demonstrating success in real-world applications will be crucial for validating its potential.

PRACTICAL IMPLEMENTATIONS

Practical implementation of QAOA requires a meticulous integration of algorithm design with available hardware capabilities and limitations. Matching problems and mixing operators to existing qubit technologies with reasonable gate depths and fidelities is imperative. Leading quantum computing companies are currently engaged in comprehensive benchmarking and performance characterization studies. Optimization of QAOA circuits and the surrounding software stack is crucial for a seamless path to deployment. Demonstrating quantum speedups over classical solvers on real applications, even with small system sizes, represents significant progress. However, scaling up meaningfully while preserving solution quality remains a formidable challenge. Continued hardware advancements in coherence, connectivity, and controls are essential to unlock QAOA's potential.

HAMILTONIAN ENCODING

One of the key steps in QAOA is encoding the problem instance into a cost Hamiltonian. The cost Hamiltonian is a quantum operator that represents the objective function of the optimization problem, such that its ground state corresponds to the optimal solution.

For example, in the case of the maximum cut problem, the cost Hamiltonian can be constructed as follows:

$H\_C = \Sigma\_{<i,j>} w\_{ij} (1 - Z\_i Z\_j) / 2$

where:
- $<i,j>$ denotes the edges of the graph
- w_{ij} is the weight of the edge between vertices i and j
- Z_i and Z_j are the Pauli-Z operators acting on qubits i and j, respectively

The cost Hamiltonian encodes the objective of maximizing the cut value, as the eigenvalue of the Hamiltonian is proportional to the number of edges between the two partitions.

Similarly, for the traveling salesman problem, the cost Hamiltonian can be constructed as follows:

$H\_C = \Sigma\_{<i,j>} d\_{ij} (1 - X\_{ij}) / 2 + \Sigma\_i (\Sigma\_j X\_{ij} - 1)^2 + \Sigma\_j (\Sigma\_i X\_{ij} - 1)^2$

where:

- <i,j> denotes the edges of the complete graph
- d_ij is the distance between cities i and j
- X_ij is the binary variable indicating whether the edge between cities i and j is included in the tour

The cost Hamiltonian encodes the objective of minimizing the total distance traveled, as well as the constraints that each city must be visited exactly once and that the tour must start and end at the same city.

The choice of Hamiltonian encoding depends on the specific problem instance and the desired trade-off between solution quality and computational complexity. In general, more complex Hamiltonian encodings can lead to better approximation ratios but may require more quantum resources and deeper circuits.

Once the cost Hamiltonian is constructed, QAOA can be used to find an approximate solution by applying a series of quantum gates that evolve the quantum state towards the ground state of the Hamiltonian. The classical optimizer is used to adjust the parameters of the quantum gates in order to minimize the expectation value of the Hamiltonian, which corresponds to minimizing the cost function of the original optimization problem.

QAOA has shown promising results on a variety of combinatorial optimization problems, demonstrating the potential of hybrid quantum-classical algorithms to provide speedups and improvements over classical algorithms. As quantum hardware and software continue to improve, the performance and scalability of QAOA are expected to increase, enabling the solution of larger and more complex problem instances.

QUANTUM-CLASSICAL HYBRID OPTIMIZATION

Quantum-classical hybrid optimization refers to the combination of quantum and classical computation to solve optimization problems. In this approach, a quantum computer is used to perform certain subroutines or steps of the optimization algorithm, while a classical computer is used to perform the remaining steps and to coordinate the overall optimization process.

The main idea behind quantum-classical hybrid optimization is to leverage the strengths of both quantum and classical computation, while mitigating their respective weaknesses. Quantum computers are good at solving certain types of problems, such as unstructured search and eigenvalue estimation, but are limited by noise, decoherence, and the difficulty of implementing arbitrary operations. Classical computers, on the other hand, are good at performing complex calculations and logic operations, but may struggle with the combinatorial complexity of certain optimization problems.

Some examples of quantum-classical hybrid optimization algorithms include:

1. Variational Quantum Eigensolvers (VQE): VQE is a hybrid algorithm for finding the lowest eigenvalue of a given Hamiltonian. It works by preparing a parameterized quantum state using a quantum circuit, measuring the expectation value of the Hamiltonian with respect to this state, and using a classical optimizer to adjust the parameters of the quantum circuit in order to minimize the expectation value.

2. Quantum Approximate Optimization Algorithm (QAOA): QAOA is a hybrid algorithm for finding approximate solutions to combinatorial optimization problems. It works by alternating between a phase of quantum evolution, governed by a problem-specific Hamiltonian, and a phase of classical optimization, which adjusts the parameters of the quantum evolution to minimize a cost function.

3. Quantum Annealing: Quantum annealing is a heuristic optimization algorithm that uses quantum fluctuations to explore the solution space of a given problem. It works by encoding the problem as a set of energy levels in a physical system, such as a superconducting qubit chip, and slowly evolving the system from an initial state to a final state that represents the optimal solution. Classical computation is used to pre-process the problem, set up the initial state, and post-process the measurement outcomes.

Quantum-classical hybrid optimization has been applied to a variety of domains, including machine learning, finance, chemistry, and logistics.

HARDWARE IMPLEMENTATION

The hardware implementation of quantum-classical hybrid optimization involves integrating quantum and classical processors into a unified system that can efficiently execute the hybrid algorithms.

One common approach is to use a quantum co-processor model, where a quantum processor is attached to a classical computer via a high-bandwidth, low-latency interface. The classical computer is responsible for executing the classical parts of the algorithm, such as pre-processing the problem, updating the parameters of the quantum circuit, and post-processing the measurement outcomes. The quantum processor is responsible for executing the quantum parts of the algorithm, such as preparing the quantum states, applying the quantum gates, and measuring the qubits.

Some examples of quantum co-processor architectures include:

1. Superconducting qubit processors: These processors use superconducting circuits to implement qubits and quantum gates. They are typically operated at cryogenic temperatures (around 20 mK) to minimize thermal noise and decoherence. Examples include the IBM Q system, the Google Sycamore processor, and the Rigetti Aspen processor.

2. Trapped ion processors: These processors use ions trapped in electromagnetic fields to implement qubits and quantum gates. They are typically operated at room temperature and can achieve high fidelity and long coherence times. Examples include the IonQ system and the Honeywell H1 processor.

3. Photonic processors: These processors use photons to implement qubits and quantum gates. They can operate at room temperature and can be integrated with classical optical networks for long-distance communication. Examples include the Xanadu X8 processor and the PsiQuantum system.

In addition to the quantum processor, the hardware implementation also requires classical control and readout electronics to manage the operation of the quantum processor and to interface with the classical computer. These include microwave and laser sources to drive the quantum gates, analog-to-digital converters to digitize the measurement outcomes, and field-programmable gate arrays (FPGAs) to implement real-time feedback and error correction.

Another approach to hardware implementation is to use a quantum annealer, which is a specialized type of quantum processor designed for solving optimization problems. Quantum annealers, such as the D-Wave system, use a network of superconducting qubits to encode the problem Hamiltonian and to perform the quantum annealing algorithm. The classical computer is used to program the quantum annealer, to read out the results, and to post-process the solutions.

SOFTWARE TOOLS

Specialized software tools are being created to support practical QAOA implementation. These include programming frameworks, simulation environments, compilers to map algorithms to hardware, optimizers to find optimal parameters and more.

Robust software to orchestrate QAOA compilation, execution, and optimization will help streamline development and maximize performance. User-friendly tools can open QAOA applications to a broader range of domain experts as well.

PERFORMANCE EVALUATION

A rigorous evaluation of QAOA performance on real and simulated quantum hardware is critical to validate its promise. Relevant metrics include solution quality, confidence bounds, convergence rates, algorithm runtime, hardware efficiency, and scaling limits.

Comparative studies against classical and quantum competitors are also informative. These benchmarks help identify areas for improvement and give clearer timelines for achieving quantum advantage in different problem domains.

SCALING UP AND LIMITATIONS

While it is theoretically possible to achieve optimality with sufficient QAOA circuit depth, doing so for large problems remains highly challenging. Key limitations centre on noise, decoherence, and exponentially growing complexity.

Pathways to scale QAOA focus on noise mitigation, problem-tailored encoding, and co-designing algorithms with hardware capabilities. Determining maximal useful problem sizes given current hardware is an important open question as well.

USE CASES AND APPLICATIONS

QAOA has broad applicability, but certain combinatorial problems are especially well-suited and impactful targets. Examples include financial portfolio optimization, job shop scheduling, power grid coordination, quantum error correction, and protein folding.

Studying performance on these domain-specific applications helps concentrate research efforts toward delivering tangible value. Partnerships with stakeholder industries and end users are thus critical.

ALGORITHM IMPROVEMENTS

Better QAOA algorithms can expand the scale and scope of problems addressable in the near term. Active research aims to enhance mixing Hamiltonians, optimize circuit depth, incorporate error suppression, and integrate problem-specific insights to improve solution quality and confidence.

Advancements in QAOA algorithms and associated hybrid software stacks will help the transition from proof-of-concept to practical tools for industry and science.

INTEGRATION WITH HYBRID ALGORITHMS

QAOA does not have to be an isolated algorithm. Combining QAOA with other quantum and classical algorithms in an overarching hybrid workflow could enhance performance. For instance, Grover's algorithm could help select good starting states or mixing Hamiltonians.

Composing QAOA with complementary techniques, both quantum and classical, offers promising directions to overcome scaling and quality limitations for certain problem classes.

TRAINING AND OPTIMIZATION

Efficiently training QAOA circuits by optimizing the mixing of Hamiltonian parameters is critical but challenging. Improved classical optimizers, surrogate models, cost function approximation, and other innovations can help accelerate training.

Tight feedback between the quantum and classical components, facilitated by fast data flows and interconnects, can also enable real-time adaptive optimization of QAOA performance.

Quantum Applications in Chemistry and Materials Science: Simulation and Discovery

QUANTUM CHEMISTRY SIMULATION

Quantum chemistry simulation strives to precisely model chemical systems and processes by directly solving the governing quantum mechanical equations for electrons and nuclei. Traditional computational chemistry methods rely on approximations, limiting their accuracy, especially for large, complex systems.

The advent of quantum computing holds the promise of enabling accurate, first-principles electronic structure simulations that are currently infeasible on classical computers. This potential revolutionizes our understanding of chemical reactions and molecular properties at a fundamental level.

Applications span drug design, catalyst development, battery materials discovery, and uncovering reaction mechanisms. Practical quantum chemistry simulations will necessitate millions of logical qubits with extremely low error rates, presenting a long-term challenge. However, even basic quantum simulations on near-term devices could offer advantages over current capabilities, profoundly accelerating progress in the chemical sciences.

QUANTUM MATERIALS SCIENCE

Quantum materials science involves leveraging quantum computing for modeling, discovery, and design of complex materials with exotic properties. This includes superconductors, topological phases, 2D materials, battery compounds, thermoelectrics, and more. Precisely simulating the quantum mechanics of vast numbers of interacting electrons and nuclei could predict material properties from first principles. This surpasses classical approximations that constrain material innovation today. Quantum algorithms could also enable high-throughput screening of hypothetical materials at unprecedented scales. In addition, quantum machine learning techniques may identify hidden patterns and guide better material designs. Quantum computing may accelerate innovations from nanomaterials to photonics to quantum technology itself. However, realizing this potential requires surmounting immense technical obstacles in developing practical quantum computers. If these challenges can be overcome, quantum materials science could take lab-based material discovery to amazing heights.

QUANTUM MACHINE LEARNING FOR CHEMICAL DISCOVERY

Quantum machine learning brings together techniques from AI and quantum computing to accelerate discoveries in chemistry and materials science. Possible approaches include using quantum neural networks and kernel methods for chemical property prediction and classification, generating molecules and materials with generative quantum models, and optimizing molecular design with quantum optimizers. These quantum techniques can potentially represent, query, and search chemical spaces exponentially faster than classical approaches. Early proof-of-concept studies on small problems exhibit promising performance improvements from hybrid quantum-classical schemes. If robustly demonstrated, quantum machine learning could become an indispensable tool for everything from drug design to the discovery of new battery materials. However, realizing practical applications requires overcoming significant technical obstacles in algorithms, software, data encoding, and quantum hardware scale-up with high reliability. Quantum machine learning for chemistry is still largely academic, but rapid advances may make it a key driver of quantum advantage in the coming decade if development challenges can be surmounted.

OUTLOOK AND CHALLENGES

Quantum chemistry and materials science stand poised to be among the earliest breakthrough application domains for quantum computing. However, many challenging obstacles remain along the path to practical realization. Important near-term priorities include developing hybrid quantum-classical algorithms and workflows to maximize early advantages, optimizing problem encoding and error mitigation techniques, and rigorous performance analysis against classical methods. Longer-term, extremely large, fault-tolerant quantum computers with high fidelities will be required to fully unlock the potential. Achieving the million-plus logical qubits needed to simulate interesting chemistry problems remains a distant milestone. Continued hardware advances, software co-design, cross-disciplinary collaboration, and application-driven research are critical to translate the tremendous possibilities of quantum computing into reality for chemical and materials sciences. Demonstrating a clear quantum advantage could catalyze broad investment and adoption among chemistry end-users and stakeholders. Despite the challenges, quantum simulation promises to fundamentally advance these fields and accelerate innovations to benefit society.

ELECTRONIC STRUCTURE METHODS

Quantum chemistry simulations rely on accurate electronic structure modelling of molecular systems. Key methods include ab initio techniques like Hartree-Fock and density functional theory. However, these make simplifying approximations. Quantum computation aims to enable exact solutions to the electronic Schrödinger equation.

This requires efficient representations of electronic wavefunctions and operators with qubits and quantum gates. Mapping desired properties into measurement observables is also crucial. If achieved, quantum simulations could provide unprecedented accuracy for chemistry and materials science.

QUANTUM PHASE ESTIMATION

Quantum phase estimation is a pivotal subroutine for quantum chemistry algorithms. It allows extracting the eigenenergies of an electronic structure Hamiltonian. This procedure applies a quantum Fourier transform to quantum states evolved under the Hamiltonian. Measurements of an ancilla register give the phase that encodes the eigenenergy. Quantum phase estimation enables extracting crucial properties for molecular modelling and material design.

EXCITED STATE ALGORITHMS

While ground states have received much focus, modelling molecular excited states is also important for chemistry. Quantum subspace expansion and variance minimization algorithms are being developed to simulate excited states.

These algorithms leverage Taylor series approximations and minimum variance principles to estimate excited state energies. Access to excited states would allow the modelling of essential photochemical processes like photosynthesis and vision.

QUANTUM MACHINE LEARNING TECHNIQUES

Multiple quantum machine learning models are being adapted for chemistry and materials applications. These include quantum neural networks for property prediction, quantum support vector machines for data classification, and quantum Boltzmann machines for generative modeling and discovery.

Hybrid quantum-classical schemes are also possible, with quantum circuits providing enhanced representations and transformations. Quantum machine learning offers additional routes to chemical insights.

QUANTUM ERROR MITIGATION

Performing useful quantum chemistry simulations will require quantum error correction and mitigation to achieve necessary fidelities and coherent evolution times. Techniques like zero noise extrapolation aim to algorithmically filter noise and reconstruct ideal outputs.

Robust error mitigation will be critical to unlocking the long-term potential of quantum simulation. Ongoing hardware advances will also help increase coherence and gate precision.

TIME EVOLUTION METHODS

Dynamic quantum simulations require modeling molecular time evolution, often through Trotterization approximations. This decomposes evolution under the full system Hamiltonian into short-time steps under individual terms.

Choosing optimal Trotter formulas and step sizes tailored to the algorithm, system, and hardware can improve modeling accuracy and efficiency. Alternative techniques like quantum signal processing are also being explored for efficient time evolution.

QUANTUM COMPUTATIONAL CHEMISTRY SOFTWARE

Specialized software stacks are emerging to support practical quantum chemistry simulation. These provide libraries of quantum algorithms, chemistry domain-specific routines, electronic structure problem encodings, hybrid quantum-classical executors, and more. Integrated, modular software environments will help chemists and material scientists leverage quantum computing without becoming quantum programmers themselves.

ALGORITHM DESIGN AND ANALYSIS

Better quantum algorithms can expand the scale and accuracy of tractable simulations. Current areas of research include more optimal phase estimation, improved Trotterization, excited states, relativistic modeling, and incorporating problem-specific structure into algorithm design. Rigorous analysis of algorithmic performance, costs, and scaling, as compared to classical techniques, helps guide R&D toward the most promising approaches..

MOLECULAR DYNAMICS

In addition to static ground state energies, modelling chemical reaction dynamics and time evolution could provide greater insights. This remains challenging for large systems. Quantum molecular dynamics algorithms based on Trotterized time evolution have been proposed. If achieved, quantum molecular dynamics could help elucidate complex reaction mechanisms and kinetics beyond the reach of experiments or classical computations.

QUANTUM SAMPLING ALGORITHMS

Quantum sampling algorithms like quantum Monte Carlo offer another route for exploring molecular design spaces and chemical discovery. The quantum advantage of quadratically faster sampling could enable broader searches. Hybrid algorithms that combine sampling with optimization and machine learning techniques may provide further synergies for chemical exploration on quantum computers.

To make robust quantum chemistry simulations practical, continued advances must focus on scale, fidelity, coherence, error mitigation, problem encoding, hybrid workflows, and more. Maintaining a rigorous application-driven mindset will help ensure these efforts produce a maximal real-world impact in chemistry, materials science, and related fields.

CHAPTER 7
PRACTICAL APPLICATIONS OF QUANTUM COMPUTING

Quantum Cryptography Protocols: Quantum Key Distribution (QKD) and Beyond

Quantum cryptography leverages principles of quantum mechanics to enable secure communication resistant to cryptanalysis. It represents one of the most advanced practical applications of quantum information science today. This section explores quantum key distribution (QKD) as the most widely implemented approach, along with post-quantum cryptography and future directions in quantum cryptography protocols.

PRINCIPLES OF QUANTUM CRYPTOGRAPHY

Quantum cryptography, also called quantum key distribution (QKD), relies on the principles of quantum mechanics to enable highly secure cryptographic key exchange between two parties. It leverages quantum properties like superposition and entanglement to detect eavesdropping attempts during key distribution, providing verifiable security, unlike classical public key cryptography. Quantum cryptography is focused on the key distribution phase rather than encryption/decryption.

The uncertainty principle and photon polarization states are foundational to quantum cryptography protocols. By encoding information in quantum states, any measurement collapses the state, so interception by an eavesdropper is detectable. This allows the communicators, typically called Alice and Bob, to determine if there has been interference, providing verifiable security.

QUANTUM KEY DISTRIBUTION (QKD) IMPLEMENTATIONS

Quantum Key Distribution (QKD) is a secure communication protocol that enables two parties to generate a shared secret key using the principles of quantum mechanics. QKD allows the parties to detect any attempt at eavesdropping, ensuring the confidentiality and integrity of the key exchange process. In this section, we will discuss the practical implementations of QKD and provide a detailed example.

QKD protocols rely on the fundamental properties of quantum states, such as the no-cloning theorem and the Heisenberg uncertainty principle, to detect any unauthorized access to the quantum channel. The most commonly used QKD protocols are the BB84 protocol and the Ekert protocol, both of which involve the exchange of quantum states (usually polarized photons) over a quantum channel and the use of classical communication to reconcile the key.

The basic steps of a QKD protocol are as follows:

1. Quantum state preparation: The sender (Alice) prepares a sequence of quantum states (e.g., polarized photons) in one of two bases (e.g., rectilinear or diagonal) and sends them to the receiver (Bob) over a quantum channel.
2. Quantum state measurement: Bob measures the received quan-

tum states in one of the two bases, chosen randomly for each state.

3. Sifting: Alice and Bob compare their bases over a classical authenticated channel and discard the measurements where their bases do not match.
4. Error estimation: Alice and Bob randomly select a subset of their remaining measurements and compare them to estimate the error rate in the quantum channel.
5. Error correction: Alice and Bob perform error correction on their remaining measurements to ensure that their keys are identical.
6. Privacy amplification: Alice and Bob apply a hash function to their keys to reduce the amount of information that an eavesdropper might have obtained during the key exchange process.

Practical QKD implementations must address several challenges, such as the generation and detection of high-quality quantum states, the synchronization and alignment of the quantum channels, and the integration with existing communication networks. Some of the most common QKD implementations include:

1. Fiber-optic QKD: This implementation uses optical fibers to transmit the quantum states between Alice and Bob. The photons are typically generated using attenuated lasers or single-photon sources and detected using single-photon detectors. Fiber-optic QKD has been demonstrated over distances of up to 500 km, but it is limited by the attenuation and dispersion of the optical fiber.
2. Free-space QKD: This implementation uses free-space optical links (e.g., telescopes and satellites) to transmit the quantum states between Alice and Bob. Free-space QKD can achieve longer distances than fiber-optic QKD, but it is more susceptible to atmospheric turbulence and background noise.
3. Chip-based QKD: This implementation uses integrated photonic circuits to generate, manipulate, and detect the quantum states on a single chip. Chip-based QKD offers the potential for low-cost, scalable, and integrable QKD systems, but it is still an emerging technology.
4. Continuous-variable QKD: This implementation uses the continuous degrees of freedom of the quantum states (e.g., the quadratures of coherent states) to encode and decode the key. Continuous-variable QKD can achieve higher key rates than di-

screte-variable QKD, but it requires more complex hardware and post-processing.

Practical Example: Fiber-Optic QKD for Secure Banking Transactions

Let's consider a practical example of using fiber-optic QKD to secure banking transactions between two branches of a bank, located in different cities (e.g., New York and Boston).

1. Setup:

    o   Alice (the New York branch) and Bob (the Boston branch) are connected by a dedicated optical fiber link, which serves as the quantum channel.

    o   Alice and Bob also have a classical authenticated communication channel (e.g., a secure telephone line) for base comparison and post-processing.

    o   Alice and Bob each have a QKD transmitter and receiver, which consist of a photon source, a polarization modulator, a single-photon detector, and a quantum random number generator.

2. Key Exchange:

    o   Alice generates a sequence of random bits using her quantum random number generator and encodes each bit into the polarization state of a single photon (e.g., 0 = horizontal, 1 = vertical).

    o   Alice sends the polarized photons to Bob over the optical fiber link.

    o   For each received photon, Bob randomly chooses to measure its polarization in either the horizontal-vertical basis or the diagonal basis (45 degrees rotated).

    o   Bob records his measurement results and the corresponding basis choices.

    o   After exchanging a pre-agreed number of photons (e.g., 1 million), Alice and Bob stop the quantum transmission.

3. Sifting:

    o   Over the classical channel, Alice and Bob compare their basis choices for each photon and discard the measurements where their bases do not match.

    o   The remaining measurements constitute the sifted key, which is typically about half the length of the raw key.

4. Error Estimation and Correction:

o   Alice and Bob randomly select a subset of the sifted key (e.g., 10%) and compare their values over the classical channel to estimate the quantum bit error rate (QBER).

o   If the QBER is below a certain threshold (e.g., 5%), they proceed with the protocol; otherwise, they abort the key exchange and start over.

o   Alice and Bob perform error correction on the remaining sifted key to ensure that their keys are identical. They can use classical error correction techniques, such as the Cascade protocol or low-density parity-check (LDPC) codes.

5. Privacy Amplification:

o   To reduce any potential information leakage to an eavesdropper, Alice and Bob apply a hash function (e.g., SHA-256) to their error-corrected keys, resulting in a shorter but more secure final key.

o   The length of the final key is determined by the secrecy capacity of the quantum channel, which depends on the QBER and the estimated information of the eavesdropper.

6. Key Usage:

o   The final key is used as a one-time pad to encrypt and decrypt sensitive banking transactions between the two branches, ensuring their confidentiality and integrity.

o   For each new transaction, a new key is generated using the QKD protocol, ensuring perfect forward secrecy.

In this example, the fiber-optic QKD implementation provides a secure and efficient way to distribute secret keys between the two bank branches, without relying on computational assumptions or trusted third parties. The QKD protocol ensures that any attempt at eavesdropping on the quantum channel will introduce detectable errors in the key exchange process, allowing Alice and Bob to abort the protocol and prevent any unauthorized access to their communication.

POST-QUANTUM CRYPTOGRAPHY

In contrast to QKD, post-quantum cryptography focuses on developing encryption algorithms from both classical and quantum computers that are resistant to cryptanalysis. This protects historical secrets and data encrypted by classical algorithms vulnerable to future decryption by quantum algorithms.

Approaches include lattice-based, hash-based, and code-based cryptographic protocols designed to be secure against attacks using Shor's algorithm and Grover's algorithm on quantum computers. Hybrid schemes combine post-quantum algorithms with existing cryptography as a transition strategy. Studies evaluate the security and efficiency of promising post-quantum cryptography candidates to standardize viable approaches before large-scale quantum computers emerge.

QUANTUM UNCERTAINTY PRINCIPLE

The quantum uncertainty principle states that complementary properties like position/momentum or polarization cannot be measured simultaneously with unlimited precision. This creates inherent uncertainty in quantum systems, collapsing superposition states upon measurement. QKD leverages this property to detect eavesdropping attempts that require intercepting and measuring photons. The uncertainty principle provides the foundation for secure quantum communication.

Quantum uncertainty enables the verifiable security of QKD protocols. By encoding information in quantum states defined by uncertain properties, measurements required for interception fundamentally alter the system in detectable ways. This allows communicating parties to determine if an adversary has interfered with the quantum channel. Uncertainty principles generally apply across quantum systems' underlying protocols using photon polarization, phase, or other quantum states.

PHOTON POLARIZATION STATES

Photons can exist in superposition states of multiple polarization bases like horizontal/vertical, diagonal, circular left/right, and more. QKD protocols encode information in these quantum polarization states. Single photons are ideal for transmission since they cannot be split to copy information. By randomly utilizing multiple polarization bases, QKD ensures an adversary cannot simultaneously measure all properties to clone the key.

Exploiting photon polarization allows QKD protocols like BB84 to distribute keys securely between two parties. By encoding information in quantum states, attempts to intercept and measure the photons collapse the superposition and are detectable. Polarization provides an ideal quantum degree of freedom for encoding cryptographic keys in optical fiber QKD systems due to its stability in transmission over long distances.

BB84 PROTOCOL

The BB84 protocol, named after its inventors Charles Bennett and Gilles Brassard, is one of the most widely used QKD protocols. It was proposed in 1984 and has since been extensively studied and implemented in various QKD systems. The BB84 protocol uses the polarization states of single photons to encode and transmit the secret key between Alice (the sender) and Bob (the receiver).

The basic steps of the BB84 protocol are as follows:

1. State Preparation:

    o   Alice generates a random bit string, which will be used as the raw key.

    o   For each bit in the raw key, Alice randomly chooses one of two bases: the rectilinear basis (0° for "0" and 90° for "1") or the diagonal basis (45° for "0" and 135° for "1").

    o   Alice encodes each bit into the polarization state of a single photon according to the chosen basis and sends the photons to Bob over a quantum channel.

2. State Measurement:

    o   For each received photon, Bob randomly chooses to measure its polarization in either the rectilinear basis or the diagonal basis.

    o   Bob records his measurement results and the corresponding basis choices.

3. Sifting:

- Over an authenticated classical channel, Alice and Bob compare their basis choices for each photon and discard the measurements where their bases do not match.
- The remaining measurements constitute the sifted key.

4. Error Estimation and Correction:

- Alice and Bob randomly select a subset of the sifted key and compare their values over the classical channel to estimate the quantum bit error rate (QBER).
- If the QBER is below a certain threshold, they proceed with the protocol; otherwise, they abort the key exchange and start over.
- Alice and Bob perform error correction on the remaining sifted key to ensure that their keys are identical.

5. Privacy Amplification:

- To reduce any potential information leakage to an eavesdropper, Alice and Bob apply a hash function to their error-corrected keys, resulting in a shorter but more secure final key.

The security of the BB84 protocol relies on the fact that any attempt by an eavesdropper (Eve) to intercept and measure the polarized photons will inevitably introduce errors in the key exchange process. This is because Eve cannot know the basis in which each photon was encoded, and any measurement in the wrong basis will destroy the photon's polarization state and introduce detectable errors.

By comparing a subset of their measurements and calculating the QBER, Alice and Bob can detect the presence of an eavesdropper and abort the protocol if necessary. The privacy amplification step further ensures that any partial information obtained by Eve is effectively removed from the final key.

The BB84 protocol has been proven to be unconditionally secure, meaning that it is secure against any attack allowed by the laws of quantum mechanics, as long as the quantum channel and the authenticated classical channel are not compromised.

COMMERCIAL QKD SYSTEMS

Prototypes in the 1990s proved the viability of QKD, leading to commercial systems becoming available in the 2000s for enterprise and government use. These provide point-to-point quantum key distribution between locations connected by fiber optic infrastructure. Keys are generated in the order of kilobits/second over distances up to around 300 km, which is suitable for one-time pad encryption of critical data.

Vendors offering commercial QKD products include ID Quantique, MagiQ Technologies, and QuintessenceLabs. Keys are distributed using dedicated fiber links or multiplexed with classical channels. Turnkey systems provide automated QKD integrated into existing IT infrastructure. Adoption is increasing, especially among defense, finance, energy, and telecom sectors requiring high-security communication.

QKD NETWORK IMPLEMENTATIONS

QKD networks extend the point-to-point QKD links to provide secure key distribution over larger distances and to multiple users. QKD networks can be classified into several categories, depending on their architecture and functionality:

1. Point-to-Point Networks:

    o   Point-to-point QKD networks consist of a single QKD link between two endpoints, such as two data centers or two branches of an organization.

    o   Example: The DARPA Quantum Network, established in 2004, was one of the first point-to-point QKD networks, connecting Harvard University, Boston University, and BBN Technologies in the Boston area.

2. Trusted Relay Networks:

    o   Trusted relay QKD networks use intermediate nodes, called trusted relays, to extend the range of QKD beyond the maximum distance of a single link.

    o   Each trusted relay acts as both a QKD endpoint and a classical relay, allowing the establishment of secure keys between distant endpoints by concatenating the keys generated on each link.

    o   Example: The Tokyo QKD Network, implemented by NEC and the National Institute of Information and Communications Technology (NICT) in Japan, is a trusted relay network that connects multiple sites in the Tokyo metropolitan area.

3. Untrusted Relay Networks:

    o   Untrusted relay QKD networks use intermediate nodes, called untrusted relays, to extend the range of QKD without requiring trust in the relays.

    o   Untrusted relays use a technique called measurement-device-independent QKD (MDI-QKD), which allows the detection of any tampering or eavesdropping by the relays.

    o   Example: The Chinese Quantum Science Satellite, launched in 2016, demonstrated an untrusted relay QKD network between the satellite and multiple ground stations, using the satellite as an untrusted relay.

4. Quantum Repeater Networks:

    o   Quantum repeater QKD networks use quantum repeaters to extend the range of QKD without the need for trusted relays.

    o   Quantum repeaters are devices that can store, process, and retransmit quantum states, enabling the establishment of end-to-end entanglement and secure key generation over long distances.

    o   Example: The European Quantum Internet Alliance (QIA) is working on the development of a pan-European quantum repeater network, aiming to connect multiple cities and countries with secure QKD links.

One notable example of a QKD network implementation is the Chinese Quantum Backbone Network, also known as the Beijing-Shanghai Trunk Line. This network, established by QuantumCTek and the Chinese Academy of Sciences, is a 2,000-kilometer fiber-optic QKD network that connects Beijing, Shanghai, Jinan, and Hefei.

The Beijing-Shanghai Trunk Line uses a trusted relay architecture, with 32 trusted relay nodes along the route. Each node is equipped with QKD devices and classical communication equipment, allowing the establishment of secure keys between adjacent nodes. The network can provide secure keys for various applications, such as secure communication, quantum secure direct communication, and quantum secure cloud computing.

The Beijing-Shanghai Trunk Line has achieved several milestones, such as:

1. A secret key rate of 47.8 kbps over a distance of 66 km between Beijing and Jinan.

2. A secret key rate of 7.9 kbps over a distance of 658 km between Beijing and Shanghai, using 11 trusted relay nodes.
3. Continuous operation for over 5,000 hours, with an average uptime of 99.8%.

The successful implementation of the Beijing-Shanghai Trunk Line demonstrates the feasibility and potential of large-scale QKD networks for secure communication and quantum-enhanced applications. However, there are still several technical and logistical challenges to overcome, such as the management and synchronization of the trusted relays, the integration with existing communication infrastructures, and the development of efficient and robust QKD protocols and devices.

LIMITATIONS AND CHALLENGES

While holding great promise, QKD has limitations, including restricted transmission distances and key generation rates suitable for distributing one-time keys rather than bulk data encryption. Security proofs depend on assumptions that real-world systems perfectly match theoretical models. Implementation challenges, such as side-channel attacks, must be addressed.

Despite these challenges, QKD, when layered with conventional encryption, provides uniquely verifiable security guarantees unmatched by classical public key cryptography. Ongoing engineering efforts aim to build on fundamental quantum security principles to make QKD a widespread reality.

HYBRID SECURITY ARCHITECTURES

Practical QKD deployments typically employ a layered hybrid architecture that integrates quantum key distribution with conventional encryption, access controls, and other information security mechanisms. QKD complements classical techniques, providing an additional layer of protection for critical communication channels.

Careful system design ensures that vulnerabilities in surrounding layers do not compromise the foundational quantum-secure keys. Hybrid systems overcome the limitations of QKD range and speed by integrating quantum-distributed keys within a broader security infrastructure. Adopting best practices and standards enables the confident construction of layered quantum-classical security architectures.

IMPROVING QKD RANGE

A major area of research aims to extend the range of QKD systems to enable long-distance quantum key distribution. Current point-to-point experiments over optical fiber and free space have demonstrated QKD exceeding 500 km. But, expanding distances further faces challenges of quantum signal loss and noise.

Approaches under investigation include using quantum repeaters for entanglement swapping, integrating QKD with telecom infrastructure, developing satellite-based quantum links, and utilizing quantum memory for stored entangled states. Improving single-photon detector technologies also aims to enable low-noise operation even with high signal loss. Overcoming range limits could eventually lead to global-scale quantum cryptographic networks.

IMPROVING QKD KEY RATES

In addition to the range, increasing the secure key generation rate remains an important goal to improve QKD viability for broader adoption. Keys are currently distributed at kilobit/second speeds, which is sufficient for one-time pad encryption but not for high-volume needs. Boosting key rates aims to support the encryption of larger data volumes.

Parallel transmission over fibre bundles or spatial-division multiplexing leverages multiple quantum channels. Wavelength division multiplexing combines quantum and classical channels on a single fibre. Faster single-photon detectors and quantum logic with lower error rates also help increase key rates. Insights from quantum information theory guide the development of protocols optimized for high-speed QKD.

REAL-WORLD QKD SECURITY

Theoretical QKD security relies on idealized assumptions. However, real-world imperfections introduce vulnerabilities that adversaries could exploit. These include side-channel attacks against physical equipment and violation of assumptions in practical implementations, allowing covert attack strategies.

Robust security analysis and testing are critical as QKD scales toward widespread use to validate security claims and uncover potential weaknesses. Certification schemes also aim to validate QKD system security. Layered architecture design and defence-in-depth strategy help mitigate vulnerabilities. Analyzing theoretical and practical QKD security helps strengthen real-world protections.

QKD STANDARDIZATION

Lack of standards hampers interoperability between different vendors' QKD systems and integration into existing infrastructure. Standardization efforts by organizations like ETSI and IEEE aim to address this by establishing norms and best practices for areas like quantum-classical interfaces, network integration, cryptographic protocols, and components.

Common standards will support trusted QKD networks with devices from multiple vendors. They aid the widespread adoption of similar historical standards that accelerated classical cryptography. Testing and certification frameworks being developed, leverage standards to validate implementation quality. Standardization provides a foundation as quantum cryptography transitions from labs into production systems.

Quantum Computing in Financial Modeling and Portfolio Optimization

Quantum computing shows immense promise for revolutionizing finance, from risk analysis to derivatives pricing to portfolio optimization. This section explores current and near-future applications of quantum algorithms for financial modelling, portfolio optimization, and other aspects of quantitative finance. Quantum techniques offer a new toolkit to transform everything from trading strategies to financial regulation.

QUANTUM ALGORITHMS FOR FINANCIAL ANALYSIS

Quantum algorithms provide techniques to tackle challenges in quantitative finance that strain conventional computing. For example, quantum Monte Carlo methods leverage quantum parallelism for complex stochastic modelling used in risk analysis and derivatives valuation. Quantum machine learning trains on financial datasets for pattern recognition, prediction, and modelling tasks.

Specific quantum algorithms offer advantages for portfolio optimization, scenario modelling, and pricing options/derivatives. Grover's algorithm speeds searches for optimal portfolios. Quantum annealing performs market simulations for stress testing and risk analysis. Quantum pricing algorithms value complex derivatives and exotic options. Ongoing research applies quantum techniques to diverse problems in quantitative finance.

PORTFOLIO OPTIMIZATION

Portfolio optimization aims to allocate assets and manage risks to maximize returns, optimize risk-adjusted returns, or meet other investment objectives. It is exponentially complex due to the vast set of options, requiring the evaluation of an immense number of possible portfolio combinations.

Quantum computing enables portfolio optimization on a scale intractable for classical techniques. Quantum algorithms search superposition states to evaluate millions of portfolio configurations in parallel. This allows for more sophisticated modelling, constraints, and risk quantification. Quantum machine learning also helps uncover hidden correlations in historical data to improve allocation decisions.

Research demonstrates quantum portfolio optimizations surpassing classical methods. As quantum processors scale, they are poised to become an indispensable tool for investment management, financial engineering, and trading strategies.

FINANCIAL RISK ANALYSIS AND MODELING

Quantifying risk is vital across banking, insurance, trading, and regulation. Quantum techniques enhance risk modelling and simulation capabilities in multiple ways. Quantum Monte Carlo methods efficiently sample probability distributions to value derivatives, simulate scenarios and evaluate risks. Quantum machine learning detects patterns and correlations in large datasets relevant to risk analysis. Quantum optimization manages complex risk models constrained by multiple nonlinear risk factors.

These exponential improvements in risk modelling, simulation, and analysis promise to boost capabilities in stress testing, capital allocation, insurance underwriting, and financial oversight. Quantum tools build on classical risk management but enable more powerful, sophisticated, and holistic risk assessment.

QUANTUM MONTE CARLO METHODS

Quantum Monte Carlo techniques leverage quantum superposition and entanglement to efficiently sample from large probabilistic spaces. This provides a speedup for Monte Carlo simulation methods used across risk analysis, derivatives valuation, and other financial applications. Quantum Monte Carlo significantly extends the scale and complexity of models by parallelizing sampling across qubit states.

Quantum Monte Carlo methods run financial simulations by preparing qubit registers into superposition states representing probability distributions of market variables. Quantum operators derived from financial models are applied to evolve the system, which is measured to collect samples. This provides an exponential advantage over classical random sampling, enabling more advanced risk quantification and valuation techniques.

QUANTUM MACHINE LEARNING FOR FINANCE

Quantum machine learning leverages quantum computation to enhance financial prediction, modelling, and pattern recognition. Applying quantum-enhanced machine learning algorithms to financial datasets can reveal correlations and insights overlooked by classical algorithms, supporting applications from algorithmic trading to fraud detection and credit risk assessment.

Techniques adapted for finance include quantum versions of neural networks, clustering algorithms, Boltzmann machines, and dimensionality reduction approaches. Hybrid quantum-classical schemes enable the processing of large financial datasets across quantum and classical hardware, opening new avenues for understanding financial markets.

QUANTUM OPTIONS PRICING

Options and derivatives pricing stands out as a key application for quantum finance. Advanced quantum algorithms have been developed to value options, swaps, and exotic derivatives. Quantum options pricing algorithms provide exponential speedups, particularly for certain classes of stochastic process models.

By running financial models on quantum hardware, these algorithms evaluate probability distributions of underlying assets. Quantum speedups result from nested Monte Carlo simulations essential for accurately valuing complex derivatives. As quantum processors scale, the possibility of real-time derivatives pricing emerges, creating opportunities in derivatives trading, hedging, and risk management.

GROVER'S ALGORITHM FOR PORTFOLIO OPTIMIZATION

Grover's quantum search algorithm brings quadratic speedup to portfolio optimization by efficiently searching through possible asset allocations. Each allocation combination is mapped to a quantum state, and Grover's algorithm facilitates quadratic speedup in finding optimal or near-optimal portfolios compared to classical brute-force searches.

This advancement allows for more sophisticated portfolio construction, meeting complex risk, return, and diversification objectives. Quantum-enhanced portfolio optimization has applications in algorithmic trading strategies, index/mutual fund management, and hedging strategies. As quantum processors scale, Grover's algorithm may become a standard tool for institutional investors and quantitative funds.

QUANTUM SIMULATIONS FOR RISK ANALYSIS

Quantum market simulations introduce a novel approach to stress testing and risk analysis across banking, insurance, and trading sectors. Quantum Monte Carlo methods efficiently run "what-if" simulations to quantify risks and dependencies, while quantum machine learning identifies patterns and generates predictive models from financial data.

These quantum tools empower firms to identify hidden risk correlations across portfolios, assess counterparty risks, and model scenarios for emerging threats and crises. Quantum techniques, when integrated with classical methods, enhance enterprise risk management and aid regulators in monitoring systemic risks arising from interconnected global markets.

QUANTUM FINANCIAL FORECASTING

Market prediction forms the foundation of quantitative finance and algorithmic trading. Quantum machine learning introduces powerful forecasting techniques by uncovering subtle patterns in noisy financial data. Quantum neural networks can model complex nonlinear relationships among economic variables, and quantum reinforcement learning maximizes trading rewards.

Quantum forecasting models have the potential to identify new predictive signals and market dynamics compared to classical algorithms. As qubit count, coherence, and connectivity improve, quantum machine learning can transform financial time-series forecasting and algorithmic trading. However, the challenge of overfitting requires careful validation.

QUANTUM ADVANTAGE VALIDATION

Demonstrating a quantum advantage is critical for financial applications. Rigorously benchmarking quantum algorithms against optimized classical alternatives helps validate real-world speedups. Areas like quantum Monte Carlo methods already enjoy asymptotic quantum supremacy for certain problem classes.

However, ongoing work seeks provable quantum advantages for broader financial use cases as hardware scales up. Hybrid schemes must balance quantum and classical workload partitioning to optimize performance. Empirical validation across diverse market data sets provides evidence for generalizable quantum speedups in finance. Achieving trustworthy quantum advantage is key for financial industry adoption.

QUANTUM PROGRAMMING FRAMEWORKS

Programming tools and algorithms tailored for finance remain limited. Development frameworks, such as IBM's Qiskit finance, strive to make quantum more accessible to finance experts without quantum skills. As hardware matures, higher-level quantum programming languages and APIs for financial applications will help lower adoption barriers.

Successful quantum finance software should hide complexity while allowing domain experts to express problems intuitively using familiar concepts from risk, portfolio theory, and derivatives pricing. Easy-to-use frameworks facilitating the mapping of financial problems to quantum circuits and algorithms will drive the utilization of quantum computing in the industry.

QUANTUM FINANCE CLOUD PLATFORMS

Cloud-based quantum computing services enable financial firms to experiment without significant upfront investment. Offerings from IBM, Amazon, Microsoft, Rigetti, and others provide access to early quantum processors and simulators, along with software tools tailored for finance. Hybrid cloud integration blends quantum and classical resources.

Cloud platforms support the investigation of quantum algorithms, the development of quantum financial models, and testing on real market data. This allows hands-on exploration to build in-house knowledge and determine which applications may become viable as hardware scales. Cloud services act as a catalyst for jumpstarting quantum finance adoption.

QUANTUM CRYPTOCURRENCIES

While not yet viable, proposals exist for "quantum money", leveraging quantum principles to create verifiably uncounterfeitable physical banknotes or digital cryptocurrencies secured by quantum effects. Approaches include public-key quantum money schemes and quantum key distribution mechanisms tailored to blockchain frameworks.

Technical hurdles remain around limited coherence times, transaction verification, and blockchain integration. But if feasible, quantum cryptocurrencies could provide advantages for minting, security, anonymity, and regulation compared to classical cryptocurrencies. Quantum techniques may provide stronger security guarantees for crypto transactions. But hype currently exceeds practical realization.

Quantum computing will likely concentrate power among financial firms able to harness the technology. This raises questions about equitable access and the potential for destabilizing markets. Cryptocurrencies and decentralized finance may provide alternative quantum-enabled models. Careful governance and cooperation between regulators, technology leaders, and financial institutions will help guide the responsible development of quantum finance.

Quantum Computing's Role in Supply Chain Management and Logistics

Managing global supply chains and logistics is a highly intricate task involving numerous variables, presenting a complex optimization and scheduling challenge. Quantum computing techniques provide a new avenue for addressing these exponentially combinatorial problems. This section explores the emerging applications of quantum computing in the optimization of supply chains, logistics, and transportation within the context of expansive global networks.

SUPPLY CHAIN NETWORK OPTIMIZATION

Modern supply chains comprise vast, interconnected distribution networks with factories, warehouses, transportation links, and retailers spanning the globe. Coordinating production, inventory, and transportation to efficiently meet demand requires optimizing resources across this vast network. The sheer complexity strains even advanced classical algorithms.

By evaluating millions of options in superposition, quantum computers can massively accelerate supply chain optimization. Researchers have developed quantum algorithms for production planning, demand forecasting, inventory management, and other processes that surpass classical techniques. As processors scale, quantum optimization promises to revolutionize global supply chain coordination.

LOGISTICS OPTIMIZATION

Logistics optimization, encompassing the management of transportation, warehousing, and delivery, encounters analogous challenges. The exponential complexity of coordinating routes, schedules, and resources across distribution networks presents significant opportunities for quantum speedups.

Quantum route optimization facilitates the rapid identification of optimal transportation routes and schedules through superposition and quantum annealing.

Quantum machine learning enhances demand forecasting and delivery anticipation using logistics data. Implementing these quantum enhancements can assist logistics providers in achieving higher efficiency and lower costs when applied across fleets, warehouses, and last-mile delivery.

DYNAMIC REPLANNING AND SCHEDULING

Supply chains must adapt to constant change - delays, new orders, equipment failures. This requires dynamic rescheduling and replanning capabilities. Combinatorial explosions make classical techniques fallible in rapidly adjusting globally optimized plans.

Quantum optimization algorithms offer a faster approach to dynamic rescheduling. As inputs change, quantum processors quickly update plans across the entire supply network. Hybrid architectures allow dynamic interfacing of quantum subroutines into overarching classical frameworks. Together, this enables resilient, optimized coordination even when disruptions occur.

PRODUCTION PLANNING

Quantum production planning algorithms optimize manufacturing operations and scheduling to enhance efficiency. By superposing possible production schedules and routes, quantum techniques identify globally optimal solutions, even for large factories with complex constraints. This capability allows manufacturers to minimize waste, energy usage, delays, and costs.

Quantum-enhanced production planning can seamlessly integrate as a subroutine in existing ERP systems, enhancing scheduling across floors, lines, machines, and material flows.

As quantum processors scale, these algorithms gain the power to dynamically optimize production in real-time, responding to new orders or disruptions. Quantum computing holds the potential to revolutionize manufacturing optimization.

INVENTORY MANAGEMENT

Inventory management across global supply chains involves enormous complexity in balancing distribution locations, stock levels, and logistics. Quantum optimization algorithms could massively improve inventory planning and deployment compared to classical heuristics.

Superposition-based quantum searches identify optimal quantities, locations, and replenishment policies to minimize costs while avoiding shortages. Quantum machine learning on inventory data uncovers predictive patterns to optimize safety stock levels. And quantum simulation models complex omni-channel distribution alternatives. With capable processors, quantum techniques hold promise for global inventory optimization.

DEMAND FORECASTING

Quantum machine learning brings powerful new capabilities to demand forecasting using purchase data, market signals, and macroeconomic indicators. Quantum neural networks can uncover complex patterns human analysts miss, better predicting customer demand.

More accurate demand forecasts enable better planning and proactive responses across manufacturing, marketing, inventory, transportation, and finance. For firms with massive training data sets, quantum techniques may significantly improve forecasting, which is critical for aligning supply with demand in dynamic markets. Hybrid classical-quantum ML approaches will likely be optimal for years until mature quantum processors emerge.

DISTRIBUTION OPTIMIZATION

Designing global distribution networks involves enormously complex optimization across manufacturing locations, warehouses, transportation modes, geography, costs, and other factors. Quantum computing enables solving such combinatorial problems at scales intractable for classical techniques.

Superposition allows the simultaneous assessment of millions of potential distribution network configurations to find optimal node locations and supply routes, maximizing efficiency and minimizing costs. But, mapping real-world distribution problems to quantum algorithms remains challenging. If achievable, quantum distribution optimization could transform global logistics topology.

TRANSPORTATION OPTIMIZATION

Scheduling transportation across land, sea, and air to deliver goods represents a complex optimization problem. Quantum computing provides new hope for enhancing route planning, load efficiency, and asset utilization across transportation modes.

Quantum route optimization considers millions of routing options in superposition to minimize costs and delays for carriers. Forecasting shipment volumes with quantum ML improves planning. For distributed sensor networks like freight telematics, quantum techniques optimize data flow. Transportation and shipping stand to benefit greatly from quantum optimization algorithms.

QUANTUM CYBERSECURITY FOR SUPPLY CHAINS

Supply chains rely on cyber-physical systems vulnerable to cyberattacks. Quantum cryptography, like QKD, provides enhanced security options for the communication networks and data underlying supply chain management. Post-quantum encryption protects sensitive operational data.

Quantum random number generators also secure sensor-based telematics systems. Quantum machine learning could help detect anomalies and cyber intrusions across supply networks. Adopting quantum techniques robust against cyberattacks will help harden the proliferation of smart supply chain systems against rising threats.

SUPPLY CHAIN SIMULATION

Today's supply chains are massively complex stochastic systems demanding advanced simulation and modelling for optimization. Quantum techniques can massively accelerate solving such probabilistic models to assess chain dynamics, risks, and disruptions.

Quantum Monte Carlo rapidly samples probability distributions governing component supply, logistics, demand, and other uncertainties. This enables faster, higher-fidelity simulation of interlinked supply chain performance to evaluate strategies under a multitude of scenarios. As with other emerging quantum applications, hybrid classical-quantum computing seems the most viable path for the foreseeable future.

SUPPLY CHAIN ONTOLOGIES

Quantum machine learning shows promise for uncovering ontological relationships from massive, interconnected supply chain datasets. Quantum neural networks could map complex correlations across products, suppliers, logistics, locations, and economic conditions.

This ontology mapping helps optimize integrated decision-making across firms. It also enables supply chain digital twins - quantum-accelerated simulations mirroring real-world systems for experimentation. With capable quantum processors, supply chain ontologies and digital twins could profoundly enhance global optimization. But for now, such applications remain theoretical rather than practical.

OPERATIONAL INTEGRATION

To provide meaningful advantages, quantum supply chain systems must connect with classical ERP, inventory, and logistics platforms. Hybrid integrations must overcome the challenges of problem encoding, data flows, and translating quantum outputs into operational decisions supported by legacy software.

Seamless integration requires the co-design of quantum and classical supply chain systems. Abstraction layers will likely emerge to ease linking enterprise platforms with embedded quantum subroutines. Platform interoperability standards can accelerate operational integration. However, realizing this vision of "quantum ERP" remains years away.

ADOPTION RISKS

Quantum computing carries adoption risks for supply chains. Limited access could overly benefit early adopters who are able to afford the technology. Integration challenges, costs, and cyber risks may deter implementation before processes mature. Workforce skills shortages may hinder organizations from effectively utilizing quantum capabilities.

Thoughtful strategies focused on incremental adoption starting from high-value subproblems can help manage risks. Governance models promoting open quantum platform access and skills development help spread benefits more equitably. Overall, though, the massive optimization potential will likely drive supply chain quantum computing adoption once it becomes viable.

OUTLOOK FOR COMMERCIALIZATION

While promising, supply chain quantum computing remains largely exploratory today. Commercialization awaits major leaps in quantum hardware development and problem mapping. But the enormous value at stake is driving R&D efforts to turn theories into practical reality over the next 5 to 10 years.

Pilot projects focused on constrained problems like shipping route optimization provide pathways to commercialization while hardware matures. Cloud quantum computing services allow further exploration without major upfront investment. Partnerships between tech and supply chain leaders will likely catalyze the development of this exceptionally disruptive innovation.

PART 3

Navigating Challenges and Future Trends

3

CHAPTER 8
CHALLENGES AND LIMITATIONS

Quantum Error Correction: Stabilizer Codes and Fault-Tolerant Quantum Computing

Quantum states are fragile and prone to errors. Quantum error correction techniques like stabilizer codes are essential to build reliable quantum computers by detecting and correcting inevitable errors. This enables fault-tolerant quantum computation that limits error propagation. There has been considerable progress in developing stabilizer codes and demonstrating elements of fault tolerance. However, substantial improvements are still needed to reach the extremely low error rates required for practical large-scale quantum computing.

BASIC PRINCIPLES OF QUANTUM ERROR CORRECTION

Quantum error correction works by encoding logical qubits in a more redundant form using multiple physical qubits. This introduces redundancy to allow detection and correction of errors without collapsing the encoded state.

The key principles are:

1. Want logical states to be unaffected by likely errors;
2. Make error-free subspaces indistinguishable;
3. Detect errors by measurements that don't disturb the encoded state;
4. Recovery procedures restore states to error-free subspaces.

Simple examples include a 3-qubit bit flip code and a 9-qubit Shor code. More complex codes provide higher fault tolerance thresholds.

Stabilizer codes are a widely used and effective approach to implement these principles. They exploit commutative measurement operators called stabilizers to detect errors without collapsing encoded states. The choice of encoding and stabilizers depends on the error models. Stabilizer codes enable scalable quantum computing if error rates are below a threshold value. However, they have high qubit overhead costs, and many codes still need to be practical. Ongoing research aims to improve encoding efficiency and develop optimized decoders.

SOURCES OF ERRORS IN QUBIT SYSTEMS

Qubit errors arise from several sources: environmental noise like thermal vibrations; control imperfections including crosstalk; manufacturing variations causing parameter differences; material defects creating decoherence centers; cosmic radiation, especially for systems without shielding; quantum measurement back-action disturbing states. Dominant errors include relaxation, dephasing, and bit/phase flips. Characterizing noise processes informs error correction strategies. Often, multiple error channels act together. Future systems will likely utilize a combination of passive error protection, active dynamical decoupling pulses, and quantum error correction.

QUBIT STABILIZER FORMALISM

Stabilizer codes detect errors through repeated measurements of multi-qubit Pauli operator combinations called stabilizers. Stabilizers have +1 eigenvalues for valid codewords and -1 for errors. Key properties: 1) Stabilizers commute with the codespace so measurements don't disturb encoded states; 2) Stabilizers commute with each other so they can be measured simultaneously; 3) Eigenvalues of stabilizers uniquely identify error syndromes. Simple examples are bit flip code detecting X errors and phase flip code detecting Z errors. Shor code detects both X and Z errors. More advanced stabilizers like surface codes provide higher fault tolerance.

STRUCTURE OF STABILIZER CODES:

Stabilizer codes have a structure consisting of the codespace of valid states, the stabilizer group generating the code, and the gauge group of operators that keep codewords within the space. Only gauge group operators change encoded information. The codespace dimension depends on qubits n, stabilizer generators m, and gauge group generators k via dimension=2n-m-k. Code efficiency depends on n,m,k overhead. Stabilizer codes can detect up to m simultaneous errors and correct up to (m-1)/2 errors. Codes are constructed to detect likely error patterns and correct important errors.

CONCATENATED QUANTUM CODES:

Concatenated codes recursively encode qubits to improve error correction strength. Inner code blocks detect some errors, while outer code detects errors in inner decoding. This builds a hierarchy of logical structure. Initial small codes are combined into larger codes in a recursive manner. Example: 3 qubit code blocks encoded in 9 qubit code, which is further block encoded. Achieves fault tolerance by limiting error propagation between levels. Enables threshold theorem, but high overhead cost. Requires consistency across code levels. Optimization of concatenated codes is an active area of research.

TOPOLOGICAL QUANTUM ERROR CORRECTION CODES:

Topological codes use lattice geometries with holes to store quantum information nonlocally, making them robust to local errors. Examples are surface codes, color codes, and defect-based codes. Stabilizers are defined from lattice geometry - measure topological properties. Provide high error tolerance thresholds in 2D. Drawbacks are high qubit overhead and complex decoding algorithms. Braiding defects allow logic operations. Demonstrated experimentally in superconducting systems. Advantages over concatenated codes but more complex decoding algorithms required.

CLASSICAL ERROR CORRECTION CONNECTIONS:

Classical and quantum error correction share common themes but also have key differences. Both exploit redundancy to protect information and use syndromes to identify errors. But quantum codes cannot copy unknown states and quantum measurements collapse superpositions. Classical codes also typically assume independent error models. Quantum error models consider correlated multi-qubit errors. Parsimonious decoding aims to find the most likely error. Classical codes inspired the development of quantum codes, but specialized quantum techniques were required.

EFFICIENT DECODING ALGORITHMS:

Decoding circuits identify and recover from errors given stabilizer measurement syndromes. Simple codes have unambiguous deterministic decoding. More complex codes require statistically inferring the most likely errors from non-unique syndromes.

Efficient classical decoding algorithms are critical to minimize time lag for recovery. Tradeoffs exist between decoder complexity, classical processing time, and achieving maximum correction power of code. Machine learning techniques may help optimize decoders. Integrating high-speed decoders with quantum hardware remains challenging. Adaptive decoding during computation is a promising area.

TRANSVERSAL LOGIC GATES:

Performing logic gates on encoded qubits Fault-tolerantly is critical. Transversal gates apply operations qubit-wise in codespace, avoiding the spread of errors. The drawback is a limited set of directly realizable gates - e.g., CNOT, Hadamard. Gate sets are not computationally universal, so ancilla protocols are required. Shor code enables a universal set but is very complex in practice. Topological braiding supports some fault-tolerant gates. Magic state distillation gives access to a broader gate set. Overall, performing universal logic within error correction codes remains highly challenging.

ANCILLA QUBIT PROTOCOLS:

Ancilla qubits enable universal logic but must be prepared in near-pure states to limit errors. A common approach is magic state distillation: repetitive parity checks filter noisy states to pure ancillas for gates. High overhead cost from distillation circuit depth. Other methods like state injection, code-switching, and gauge fixing are proposed to reduce overhead. Integrating optimized ancilla protocols with stabilizer codes while maintaining a fault tolerance threshold is an active research target.

THE QUANTUM THRESHOLD THEOREM

The threshold theorem demonstrates that if the error rate per operation is below a threshold value, scalable quantum computing is possible through stabilizer codes. The threshold depends on code parameters, decoder efficiency, gate types, and error models. Early estimates ranged from 10-3 to 10-6 but have improved over time. Recent studies suggest the threshold could be between 0.1% to 1%. Reducing physical error rates is thus one of the most critical areas of focus. Reaching very low logical error rates will require a combination of approaches..

QUANTUM MEMORY AND REPEATER CODES

Specialized codes aim to just preserve quantum data over time. Simple repetition codes repeatedly measure/refresh qubits. Surface codes offer high storage thresholds with 2D arrays. Topological codes have promise for robust long-term storage. Codes optimized just for memory have lower resource overhead than computation codes. Linked quantum repeater networks would enable long-distance transmission of quantum information encoded across the links. This requires good storage codes and fault-tolerant swap operations.

CODE TRANSFORMATIONS

Transforming between different quantum error correction codes enables tailoring them for specific purposes. For example, start with an efficient encoding circuit but transform it to a better code later. Or begin with a limited connectivity surface code and then transform it into 3D colour codes. Code deformations like lattice surgery enable these transitions. Augmenting codes by increasing encoding is also possible. Intuitively, quantum information fluidly flows between changing code structures. Developing optimized transitions between codes to balance overhead costs, error tolerance, and desired operations is an open challenge.

GAUGE FIXING PROTOCOLS

Stabilizer codes have equivalent states related to "gauge" transformations. Gauge fixing periodically rotates code to a known frame as errors accumulate. This limits error spread but adds the cost of gauge circuits. Examples are twirling, which rotates arbitrary errors to Z-basis errors, and code resetting circuits. Gauge protocols also enable optimized ancilla use and help control decoherence in logical qubits. When combined with parity checking, it allows continuous correction. Adds operational complexity but helps maintain code space.

SYNDROME HISTORY RECONSTRUCTION

Stabilizer codes assume memoryless errors, but realistic noise has correlations over time. To address this, syndrome history methods consider correlations in past measurements to better infer the most likely current errors. However, long histories exponentially grow decoding complexity. Efficient algorithms use neural networks, Monte Carlo models, and belief propagation to approximate optimal Bayesian decoding. History reconstruction improves accuracy, but classical processing must keep pace with quantum systems. Also requires efficient fault-tolerant readout of records of past syndromes.

BLIND QUANTUM COMPUTATION

Blind quantum computing enables users to delegate computation to a remote quantum server without revealing inputs. The server measures stabilizers and returns limited classical data for the user to reconstruct the result. It enables cloud quantum computing but requires the server to perform remote error correction protocols, which limits practicality initially. Blind computing security proofs rely on fault-tolerant encodings. It may have applications in secure distributed quantum networks. Allows broader access to early quantum processors.

DYNAMIC QUANTUM ERROR CORRECTION

Most codes assume perfect gates - limitations for NISQ devices. Dynamical decoding adapts in real-time to gate errors revealed through measurements, avoiding flawed assumptions. Feedback optimizes decoding by correlating observed errors with gates. But decoding complexity grows exponentially with circuit depth. Efficient inference of most likely errors dynamically is challenging, especially for temporally correlated non-Markovian noise. Active area of research but ultimately limited coordination of classical and quantum systems needed.

MACHINE LEARNING FOR ERROR CORRECTION

Machine learning techniques like neural networks are gaining traction for optimizing error correction performance. It can help speed up decoding and improve accuracy by recognizing subtle correlations in stabilizer outcomes. Reinforcement learning allows customizing codes for given hardware by exploring recovery strategies through trial and error. Classifiers identify likely error sources from signatures. Autoencoders reconstruct quantum information from limited measurements. Significant potential, but run-time demands of decoding circuits remain a bottleneck.

TRADEOFFS BETWEEN CODE OVERHEAD AND HARDWARE QUALITY

Error correction imposes large resource overheads that must be balanced with raw qubit quality. Codes with lower overheads typically tolerate less noise. Surface codes offer balanced performance but high overheads to reach fault tolerance. The major outstanding question is the crossover point where improving hardware alone is more beneficial than elaborate error correction. Overall, the solution will use code overhead minimally needed for given hardware. This depends on future improvements in materials, manufacturing, controls, and architecture.

Small demonstrations of stabilizer encoding, error detection, and correction have been achieved in systems like superconducting qubits and trapped ions. But these are far from practical applications. Current experiments focus on validating elements like repeated syndrome measurements, syndrome-based recovery, and enhanced logical qubit lifetimes. Implementing complete end-to-end encoding, decoding, gate operations, and fault tolerance remains highly challenging. Ongoing work focuses on improving gates, integrating decoders on-chip, developing optimized codes, and benchmarking.

Scalability Challenges: Quantum Volume and Quantum Supremacy Milestones

Practical quantum computing's realization hinges on the progression to systems boasting millions of qubits and achieving remarkably low error rates. While benchmarks like Quantum Volume and Quantum Supremacy gauge progress, meaningful scaling remains an immensely challenging endeavour. The hurdles encompass expanding qubit numbers, minimizing errors, developing robust software stacks, identifying useful applications, and orchestrating advances across diverse disciplines and institutions. It is crucial to surpass the limitations of today's noisy intermediate-scale quantum (NISQ) devices to unlock the transformative potential of quantum computing.

EXPANDING QUBIT ARRAYS:

All qubit technologies face challenges in expanding qubit numbers while maintaining control fidelity and connectivity. For superconducting circuits, this involves integrating more qubits on a chip while minimizing crosstalk. Trapped ions must add trap zones and control electrodes without excess micromotion. Cold atoms require larger optical lattices and more laser control beams. Photonics needs more efficient generators, detectors, and interferometers. Hybrid systems aim to combine multiple qubit types but have interface challenges. Significantly enhanced control electronics will also be necessary to address larger qubit arrays.

PRESERVING QUBIT COHERENCE:

Larger systems make detailed individual qubit performance characterization more difficult. However, maintaining long coherence times across all qubits will require improvements in material synthesis, substrates, fabrication techniques, and shielding. This demands research advances across physics, materials science, and engineering domains. Tradeoffs exist between directly connecting qubits versus networked modular architectures. The optimal configuration will depend on the target application and underlying qubit technology capabilities.

MINIMIZING CROSSTALK AND CONTROL ERRORS:

Increasing qubit numbers risks more unintended interactions that reduce control fidelity. Methods are needed to better characterize and suppress crosstalk across large arrays. Control electronics must also scale up and operate faster without injecting more noise. Pulse shaping techniques counter bandwidth limits and distortions. Modular architectures can isolate subsystems, while quantum error correction limits the propagation of control errors. Maintaining high-fidelity operations as systems grow will require coordinated progress across electronics, materials, and software.

BENCHMARKING QUANTUM VOLUME:

Quantum Volume is an industry benchmark measuring system performance. It combines qubit number, connectivity, and error rates into one metric, reflecting the largest random circuit that can be implemented successfully. High Quantum Volume requires scaling up qubit arrays and reducing gate errors. The current state-of-the-art is Quantum Volume 64, but practical applications likely require millions of qubits. Quantum Volume guides hardware development priorities for the noisy intermediate-scale quantum (NISQ) era. Useful Quantum Volume tests must also be efficiently verifiable on real hardware.

CHARACTERIZING QUANTUM CHAOS AND COMPUTATIONAL FIDELITY:

Quantum chaos studies how small perturbations affect system evolution. This provides insights into error accumulation and limits to computable circuit depth. Measuring quantities like butterfly velocities and out-of-time-order correlations reveals coherence decay rates. Quantum chaos sets bound on usable coherence before algorithm-level error correction is needed. Detailed noise modeling for target problems will allow for the co-designing of robust algorithms and tailored error correction strategies.

CLASSIFYING QUANTUM ERROR PROCESSES:

Detailed error models considering non-Markovian noise processes are essential for optimizing error correction. Measurement techniques like randomized benchmarking, gate set tomography, and detector tomography provide insights into error types, locations, and likelihoods.

Machine learning techniques help identify correlations and patterns in complex system dynamics. Maintaining up-to-date error models as systems grow poses scaling challenges. Efficient validation of noise processes across large arrays will become critical.

REALIZING LOGICAL QUBITS:

Logical qubits built from error correction codes are key to scalable quantum computing but extremely challenging to implement. This requires full integration of encoding, decoding, gate operations, and error correction protocols while maintaining fault tolerance.

Current experiments are primitive demonstrations only. Major outstanding challenges include reducing overhead, increasing code distance, implementing fast decoder circuits, and demonstrating scalability to multiple logical qubits. Logical qubit demos would boost confidence in long-term prospects of practical error correction.

EMBRACING MODULARITY AND HETEROGENEITY:

Modular architectures allow the combining of multiple qubit types and co-processors tailored to specific tasks, like quantum annealing units or photon sources. This heterogeneous approach aims to utilize the strengths of each technology.

But, it imposes big challenges in connecting systems while limiting noise contamination. Hybrid systems also enable exploring qubit interfaces like microwave-to-optical transducers. But, major engineering advances are still required to build controllable, scalable hybrid platforms.

INTERFACING CLASSICAL AND QUANTUM SYSTEMS:

Practical quantum computing requires rapid interfacing between classical and quantum hardware at scale. This involves swift data conversion, storage, and processing, with multiplexing across many control lines essential for configuring and readout of large qubit arrays. On-chip controllers reduce latencies but pose fabrication challenges and introduce complex cryogenic design constraints. The integration of smart streaming algorithms and feedback systems aims to minimize classical quantum bottlenecks.

DEVISING PRACTICAL ERROR CORRECTION STRATEGIES:

Theory provides many error correction options, but practical implementations must balance performance and complexity. This requires a holistic understanding of hardware capabilities and limitations to guide code selection. For example, surface codes offer good thresholds with 2D layouts, while color codes promise high performance requiring 3D connectivity. Practical codes will likely utilize a combination of approaches tailored to hardware constraints. Teams of experts across physics, materials, computer science, and engineering are needed.

REDUCING THE RESOURCE OVERHEAD:

Current error correction schemes exhibit substantial resource overheads, necessitating cost reduction through optimized codes, efficient decoders, advanced gate techniques, and hardware improvements. Adapting codes dynamically based on observed error rates and models can lower overheads.

Reducing gate times through materials advances and coherent control techniques is beneficial. Striking a balance between overhead minimization and error correction power remains key. Real applications will utilize the necessary overhead for a given hardware quality.

DESIGNING QUANTUM MEMORY UNITS

Specialized quantum memory units are attractive for storing and relaying quantum information within future large-scale computing architectures. Memory enables on-demand retrieval of qubits for tasks like quantum repeaters. This relieves constraints on direct qubit connectivity. Codes optimized specifically for storage have lower resource overhead than computation codes. However, integrating high-performance memories with processors represents a major engineering challenge, especially in cryogenic implementations.

REALIZING A UNIVERSAL GATE SET

A universal gate set is necessary for general quantum computation. Many proposed gate sets are not inherently fault-tolerant or have limited direct realizations in error correction codes. Ancilla protocols enable universality but add significant complexity. Topological braiding supports a universal gate set but needs to be more practical. Discovering new gate sets tailored to code constraints and hardware capabilities would be enabling. But for now, surface code schemes provide the most promising path by combining magic state distillation with code operations.

DEVELOPING ROBUST QUANTUM SOFTWARE STACKS

Efficient software is critical to utilizing NISQ hardware and unlocking the potential of future fault-tolerant devices. This spans low-level controls to high-level application interfaces. Significant work remains to enable robust compilation, optimization, error mitigation, and control workflows. Architecture co-design balancing hardware capabilities and algorithm needs will be essential. Modular software stacks integrating diverse modules like simulators, debuggers, and compilers remain under development across the industry.

FINDING COMMERCIALLY VALUABLE APPLICATIONS

Applications that demonstrate a quantum advantage in the NISQ era still need to be determined. Leading candidates are optimization, quantum chemistry, and quantum machine learning. However, defining commercially viable applications robust to imperfect qubits is critical to focus quantum computer development. Areas like quantum simulation may eventually be powerful but require millions of logical qubits. Defining a realistic incremental roadmap for delivering quantum enhancement in useful domains will build confidence and provide feedback to hardware developers.

ADDRESSING THE CRYOGENIC CHALLENGE

Superconducting and several other qubit technologies require complex cryogenic systems with multiple temperature stages and shields. Engineering reliable, user-friendly systems that can scale to large footprints represents a major challenge. Advances in cryocooler technologies, cabling, and thermal management are needed. Creative solutions like cryo-CMOS circuits help, but heat dissipation constraints remain. The cryogenic challenge is a critical bottleneck for scaling up these quantum hardware platforms.

INVESTING IN QUANTUM ENGINEERING

Designing, building, optimizing, and operating practical quantum computers requires cultivating a skilled workforce of quantum engineers. These interdisciplinary experts straddle computing, materials science, control theory, electronics, firmware, and software engineering domains. Quantum engineering must emerge as a distinct discipline combining broad technical expertise with specialized quantum skills. Training programs at universities and companies are growing but must be expanded to match industry growth.

ATTAINING PUBLIC AND PRIVATE INVESTMENT

Advancing quantum computing to commercial viability requires massive investment from both the public and private sectors. While funding is growing significantly, sustaining commitment through both technical successes and inevitable setbacks will be key. Near-term government funding should target high-risk basic research and infrastructure development. The private sector can accelerate commercialization, but sustainable markets are still developing. Maintaining momentum across sectors, institutions, and countries is critical.

EXPANDING INTERNATIONAL COLLABORATION:

Advancing the field of quantum computing necessitates a global effort, requiring collaboration among leading institutions in academia, government labs, and corporations spanning numerous nations. Recognizing that no single country can dominate across all domains, effective collaboration becomes imperative. Overcoming challenges tied to political tensions and intellectual property constraints is essential to drive rapid progress. Successful international partnerships, exemplified by initiatives like QED-C, focused on demonstrating error correction, serve as models for achieving collaborative success.

SETTING REALISTIC EXPECTATIONS:

The current landscape of quantum computing is marked by tremendous excitement; however, expectations must align with realistic technology forecasting. While algorithm research can operate under the assumption of ideal fault tolerance, hardware development must proceed incrementally, acknowledging the inherent complexities. Overly optimistic predictions regarding development timelines risk subsequent disillusionment. The quantum computing community needs to provide cautious roadmaps that reflect the immense technical challenges lying ahead. Although quantum advantage in specific applications may materialize within 5-10 years, the realization of universal fault-tolerant quantum computing likely remains decades away.

Overcoming Decoherence: Error Rates and Error Sources

Decoherence stemming from environmental noise and control imperfections leads to qubit errors and the loss of quantum information. To effectively reduce error rates, a multifaceted approach is required.

This includes advancements in materials, the implementation of smarter pulse sequences, operating qubits at their optimal sweet spots, and utilizing quantum error correction techniques. Further research is also essential to comprehensively characterize the diverse error sources and noise processes that impact various qubit technologies. This understanding will guide the development of strategies aimed at combating decoherence.

ENVIRONMENTAL NOISE SOURCES

Qubits are sensitive to noise fluctuations in the environment that cause decoherence. Examples include vibrational, acoustic, and thermal noise; electromagnetic interference; defects in materials and at interfaces; crosstalk between control lines; and cosmic radiation. Trapped ion qubits are limited by motional heating. Superconducting qubits are affected by two-level system defects. Different qubit technologies will have distinct noise sensitivities that should guide materials and architecture choices.

CONTROL ELECTRONICS IMPERFECTIONS

Control electronics inject noise if gain, bandwidth, and crosstalk are not carefully managed. 1/f noise from unstable bias sources and thermal noise from attenuators degrade fidelity. Pulse shape distortions and rise time limitations cause gate errors. Electromagnetic interference leaks in through interfaces. As systems scale up, fast, low-noise control hardware is critical. Careful engineering combined with pulse shaping techniques can mitigate electronics limitations.

MANAGING QUBIT CROSSTALK

Unintended qubit interactions from crosstalk must be minimized, especially as systems grow larger. Superconducting qubits suffer ZZ and ZX crosstalk, mediated by resonator bus or chip substrate. Methods to better isolate qubits like 3D cavities, tunable couplers, and all-microwave controls help to decouple neighbours. Trapped ions utilize optic beam focusing and shuttling sequences to suppress crosstalk during multi-qubit gates. Architectures should aim to minimize crosstalk channels.

QUANTUM MEASUREMENT ERROR PROCESSES

Quantum measurements can disturb qubit states or have technical errors that degrade fidelity. QND measurements aim to overcome back-action but are challenging to implement. Dark counts, discrimination errors, and non-unit efficiency affect detectors. Time resolution limitations cause errors. Some technologies have better fidelities than others for readout.

Fast, high-fidelity single-shot measurements are critical for error correction but difficult to scale up. Reducing measurement errors remains an active area.

MANUFACTURING AND MATERIALS IMPROVEMENTS

Qubit's performance is limited by defects and imperfections in materials and fabrication processes. Enhancing material purity and crystalline quality reduces decoherence centers. Optimizing surface quality and qubit-substrate interfaces is also beneficial. Superconducting qubits benefit from improved dielectrics for Josephson junctions. Better connectivity in 3D qubit arrays necessitates advanced nanofabrication. Achieving atomic-level precision control over manufacturing will help suppress errors.

LEVERAGING QUANTUM CHAOS STUDIES

Quantum chaos examines how perturbations affect system dynamics. This reveals intrinsic error tolerances and decoherence mechanisms. Analyzing quantities like entanglement growth, scrambling time, and out-of-time-order correlators exposes error accumulation rates. These fundamental bounds guide the development of robust control methods and tailored error correction strategies. Quantum chaos illuminates pathways and pitfalls towards quantum advantage in noisy devices.

SWEET SPOTS AND OPTIMAL BIAS POINTS

Qubits often have sweet spot operating points where they are naturally less prone to decoherence. Examples include zero-first-order Zeeman shifts in trapped ions and flux noise-insensitive spots in superconducting qubits. But architecture should not rely solely on sweet spots. Understanding and calibrating optimal bias points reduces sensitivity to noise. Combining sweet spot operation with active pulse techniques provides in-depth defence against errors.

COHERENT PULSE CONTROL TECHNIQUES

Applying carefully tailored pulse sequences based on coherent control principles can help suppress coupling to noise modes and counteract systematic errors. Precisely shaped pulses steer qubits along optimized trajectories, avoiding pitfalls. Dynamical decoupling sequences prolong coherence times. Walsh modulation generates robust arbitrary gates. These active techniques provide powerful knobs for error mitigation independent of underlying materials or technology.

HAMILTONIAN ENGINEERING APPROACHES

Tuning qubit interactions can reduce sensitivity to noise. For example, modifying superconducting qubit formulas changes noise coupling. Synthetic clock transitions decouple trapped ion qubits from magnetic field fluctuations. Qubit connectivity can also be tuned to minimize crosstalk. Hybrid systems combine advantageous noise properties of different qubit types. Hamiltonian engineering should be informed by noise characterization to mitigate dominant error channels strategically.

LEVERAGING QUANTUM ERROR CORRECTION

Quantum error correction provides a general strategy for overcoming decoherence by encoding logical qubits redundantly. This enables scalable quantum computing if physical error rates are below a threshold value. While significant overhead is required, combining error correction with other techniques reduces demands on each. Even partial error correction can enhance coherence relative to bare physical qubits. Developing low-overhead codes tailored to hardware will help bridge to full fault tolerance.

SPECIALIZED LAB PROTOCOLS

Specialized lab procedures can enhance qubit performance by mitigating environmental noise. Examples include high-purity vacuum systems, vibration isolation slabs, RF and magnetic shielding, and battery-powered electronics. While not feasible in final packaged systems, these techniques improve testbed fidelities and help identify noise processes. They also allow exploring physics limits of coherence absent technical noise. Standardized protocols for qubit characterization will aid comparability.

SYSTEM DESIGN FOR NOISE ISOLATION

System and circuit design techniques should aim to isolate qubits from noise sources. Examples include placing qubits in ground planes to limit substrate coupling and using superconducting vias for crosstalk shielding. Careful component layout minimizes parasitic modes. Optimized wiring geometries reduce crosstalk. Such careful engineering design approaches will become even more critical as systems scale up to large numbers of qubits.

QUANTUM NON-DEMOLITION QUBIT READOUT

Quantum non-demolition (QND) measurement techniques enable high-fidelity qubit readout without disturbing the qubit state. This reduces back-action errors but is challenging to implement. Typical approaches involve ancilla couplings or parametric modulation. When combined with high-efficiency detection, QND readout gives a major boost in measurement fidelity. This is especially beneficial for repeated error syndrome measurements in quantum error correction codes.

REAL-TIME FEEDBACK AND ADAPTIVE CONTROL

Actively monitoring qubits and providing real-time feedback enables adaptive control to counteract noise and errors. This allows for optimizing qubit settings, compensating for drifts, and customizing pulses. However, latencies from classical processing limit speed. On-chip controllers allow faster response but add complex cryogenic design constraints. Developing smart adaptive techniques that work within feedback bandwidth constraints poses interesting challenges.

CHARACTERIZING NOISE PROCESSES THROUGH MODELS

Precise noise models considering non-Markovian and non-Gaussian noise processes inform strategies to optimize error suppression. Measurement techniques like randomized benchmarking elucidate noise characteristics. Physical models examine sources like 1/f charge noise or phonon decays. Machine learning assists in identifying noise correlations and signatures. Detailed error analysis illuminates pathways for improving materials, designs, and controls to enhance coherence.

VALIDATING AND UPDATING SYSTEM MODELS

Maintaining accurate error models as systems scale poses significant challenges. Efficient techniques for full system benchmarking will be essential. Statistical learning methods can help detect model drift over time. Error models may have complex dependencies on operating points, necessitating multi-parameter characterizations. Automated testing and calibration routines will help keep models aligned with hardware as it evolves. This enables customized error mitigation strategies.

DESIGNING ROBUST PULSE SEQUENCES

Carefully tailored pulse sequences can steer qubits along precise trajectories to minimize sensitivity to noise and avoid pitfalls leading to decoherence. Pulse shape, timing, and modulation can be optimized to counteract dominant error processes revealed through models. Sequences can also be made resilient to calibration errors and distortions. Co-designing robust sequences with accurate error models will become increasingly critical for quality control as systems scale up.

MANAGING QUBIT VARIABILITY

There are always small variations across qubits in a system due to manufacturing precision limits. Characterizing and then compensating for this variability through custom controls and tuning helps improve uniformity and reliability. Adaptive control techniques also dynamically calibrate to qubit-specific parameters. Gate error rates improve if controls are customized for qubit variations rather than one-size-fits-all approaches. This tuning process becomes more challenging as qubit numbers increase.

LEVERAGING QUANTUM ERROR MITIGATION

Quantum error mitigation aims to reduce noise impact by estimating its effect on algorithm output and counteracting through modifications like parameterized noise injection or inversion. This provides partial protection that is more accessible than full error correction. However, accurate noise models are required. Mitigation also introduces overheads from additional characterization, gates, and classical processing. Balancing costs and fidelity improvements remains challenging.

TARGETING REMAINING DECOHERENCE SOURCES

As known sources are gradually suppressed through advances in materials, design, and control, remaining decoherence processes become apparent. These typically involve subtle, complex effects like non-Markovian noise correlations.

Eliminating these residual errors likely requires a coordinated strategy leveraging optimized pulse techniques, error correction, duty-cycle management, and high-precision calibration.

A deep understanding of quantum dynamics in engineered systems is needed to eradicate the final imperfections.

CHAPTER 9
QUANTUM COMPUTING ETHICS AND GOVERNANCE

Quantum Information Security: Post-Quantum Cryptography and Quantum-Safe Solutions

One of the most pressing governance challenges raised by quantum computing is the threat it poses to current standards of information security, especially public key cryptography. Quantum algorithms like Shor's can efficiently break the encryption schemes commonly used to secure data today. To safeguard information in the emerging quantum era, new post-quantum cryptographic solutions and frameworks will be urgently needed. However, this transition also surfaces myriad complex technical, economic, and coordination obstacles. Navigating the migration to post-quantum cryptography securely and responsibly will require concerted efforts between policymakers, regulators, industry leaders, technology standards bodies, and academia.

UNDERSTANDING THE THREAT QUANTUM COMPUTERS POSE TO EXISTING CRYPTOSYSTEMS

Shor's Algorithm and the Vulnerability of RSA Encryption

Shor's quantum algorithm for integer factorization, introduced by MIT's Peter Shor in 1994, challenged prior assumptions about computational complexity and public key cryptography. By leveraging principles of quantum superposition, Shor's algorithm can factor large integers exponentially faster than any classical counterpart. This poses a substantial threat to RSA encryption, where security hinges on the assumed intractability of factoring large prime numbers.

Shor's algorithm transforms this traditionally intractable problem into one highly tractable on quantum computers. This groundbreaking development reverberated through the cryptography community, revealing that quantum computers could essentially "break" a significant portion of modern public key cryptography.

Implications for Widely Used Cryptography Standards and Protocols

The vulnerability of RSA encryption to quantum attack via Shor's algorithm has profound implications for the various security protocols and cryptographic standards built off RSA. Transport Layer Security (TLS) and Secure Sockets Layer (SSL), used to secure internet communications, rely critically on RSA and other vulnerable public key schemes. Standards like Diffie-Hellman key exchange and Digital Signature Algorithm are also threatened in the post-quantum era. Major asymmetric cryptography frameworks trusted today for encrypting sensitive data, authenticating users, and guaranteeing integrity are fundamentally jeopardized by the ability of quantum computers to undo their core mathematical assumptions. New post-quantum secure solutions are needed.

POST-QUANTUM CRYPTOGRAPHIC STANDARDS AND PROTOCOLS

Leading post-quantum cryptography schemes and their properties

To safeguard systems against cryptanalysis from quantum computers, new standards and protocols for post-quantum or quantum-safe cryptography must replace today's vulnerable public key-based systems. The primary post-quantum schemes encompass lattice-based cryptography, code-based cryptography, multivariate polynomial cryptography, hash-based signatures, and symmetric key quantum cryptography.

Each approach relies on distinct mathematical assumptions outside the efficient reach of Shor's algorithm and other known quantum attacks. The various schemes have strengths and weaknesses in areas such as performance, key sizes, and susceptibility to classical brute force attacks. Evaluating these trade-offs will guide the selection of the most suitable standards for different applications.

Efforts by NIST and Other Standards Bodies to Evaluate and Standardize Post-Quantum Cryptography

Recognizing the need to transition to post-quantum security, standards bodies like the National Institute of Standards and Technology (NIST) have been evaluating candidates and working to standardize new quantum-safe cryptosystems. NIST has winnowed down initial proposals to a small number of finalist algorithms across encryption, signatures, and key exchange. Through its post-quantum cryptography standardization project, NIST aims to provide guidance to organizations and software developers on selecting the most practical and secure quantum-resistant schemes.

Other standards groups like the IETF and ISO are also working to update their own standards and best practices to the post-quantum era. These efforts will facilitate the ecosystem-wide adoption of standardized post-quantum cryptography.

REAL-WORLD DEPLOYMENT OF POST-QUANTUM CRYPTOGRAPHIC SYSTEMS

Technical challenges and vulnerabilities in transitioning to post-quantum cryptography

The real-world deployment of standardized post-quantum cryptographic systems poses major technical hurdles. Software and hardware broadly need to be upgraded to integrate with new quantum-safe algorithms. The possibility of flawed implementations leaving systems vulnerable during transition must be guarded against. Timing also presents challenges - data needs protection before large-scale upgrades can be completed, requiring hybrid old and new systems. Execution differences between classical and quantum-resistant cryptographic operations may introduce new attack surfaces like side-channel leaks. Thoughtful engineering and conservative rollouts will be imperative for managing these technical complexities.

Costs and coordination challenges of widespread post-quantum cryptographic adoption

In addition to technical obstacles, the worldwide upgrade to post-quantum cryptography represents a massive economic and coordination challenge. The costs of transition will be substantial across industries and governments. Required upgrades span hardware, software, communications infrastructure, embedded devices, control systems, and more. But the risks of failing to upgrade are far higher. Distributing costs and aligning efforts by sector, public-private partnerships, and industry consortiums will be critical. Policymaker assistance and staged timelines can also smooth obstacles. Close global coordination is essential for navigating this disruption.

QUANTUM KEY DISTRIBUTION FOR QUANTUM-SECURE COMMUNICATIONS

Properties of quantum key distribution and quantum cryptography systems

In addition to post-quantum cryptography based on computational complexity, quantum key distribution (QKD) offers another path to quantum-secured communications. QKD uses fundamental properties of quantum states like uncertainty and entanglement to generate and distribute symmetric encryption keys securely. Unlike post-quantum cryptography, QKD's security stems from the laws of physics rather than mathematical difficulty. This provides "unconditional" security guaranteed by physics against adversaries like quantum computers. However, QKD has limitations around transmission distance, key generation rate, and cost compared to post-quantum cryptography.

QKD's advantages and limitations for real-world deployment

While conceptually appealing, QKD faces substantial practical obstacles to large-scale deployment. Its properties mean QKD is best suited for specialized applications like high-value secure links between physically proximate nodes rather than internet-scale communications security. Properly engineered quantum networks may eventually widen applications but will require major infrastructure investment. For most applications, post-quantum cryptography is likely to be more flexible and cost-effective once standards mature. But for niche security needs, QKD provides unique quantum-based guarantees.

TRANSITIONING ENCRYPTED DATA TO POST-QUANTUM SECURE SCHEMES

Techniques and risks involved in re-encrypting existing data

One of the most fraught challenges in migrating to post-quantum cryptography is developing transition plans and protocols for re-encrypting or destroying existing encrypted data using vulnerable schemes. Some data may need re-encryption, while other types may require destruction if risks are too high should quantum brute forcing become possible. However, at scale, re-encryption can be extremely resource-intensive.

Data destruction also inevitably incurs a loss of information. Estimating risks and costs will require categorizing data types, analyzing dependencies, and developing heuristics. Standards bodies and policymakers have roles in setting best practices for these processes.

Handling encrypted data in archived and backup systems

Special care must also be taken to handle encrypted data persisted in archived systems and backups during the quantum transition. This data still needs protection, either through re-encryption or destruction. But, archived and backup data often need to catch up to active systems in updates. Policies will need to determine acceptable timelines for remediating old encrypted data based on estimated quantum threat horizons. More complex re-encryption schemes leveraging both post-quantum and traditional algorithms may allow some archived encrypted data to be kept past threshold dates.

Ethical Considerations in Quantum Computing: Privacy, Bias, and Fairness

In addition to information security concerns, quantum computing raises broader ethical issues around privacy violations, algorithmic biases, and overall fairness. Quantum techniques for optimization, machine learning, and cryptanalysis could exacerbate existing data mining and surveillance risks. Complex quantum machine learning systems may also amplify and obscure unfair biases. To uphold principles of responsible scientific development and human dignity, technologists and policymakers should proactively engage with these emerging quantum ethics challenges through technical safeguards, governance frameworks, and public dialogue.

PRIVACY RISKS AND SAFEGUARDS IN THE ERA OF QUANTUM COMPUTING

Potential risks to personal privacy from quantum data analysis capabilities

By enabling new capabilities in pattern finding, prediction, and decryption, quantum computing could heighten existing threats to personal privacy. Quantum machine learning algorithms may amplify risks around profiling individuals based on data like purchases, location, biometrics, and browsing history. Their optimization capacities could also aid in deanonymizing people hidden in datasets. Additionally, quantum cryptanalysis could undo previous techniques for protecting personal data. These emerging capabilities warrant strengthening existing technical privacy safeguards and governance protections.

Technical and policy mechanisms to mitigate privacy risks from quantum computing

To counter privacy risks exacerbated by quantum computing, both technical and governance safeguards will be important. Technically, approaches like differential privacy, homomorphic encryption, and multi-party computation may gain wider use to prevent the exposure of personal data during quantum computation. Governance-wise, developing clear regulations and international norms around the use of quantum techniques for surveillance, informed consent in quantum AI systems, and limitations on decrypting certain classes of data will help establish guardrails. Public engagement and ethics boards also have roles in providing guidance.

RECOGNIZING AND MITIGATING BIASES IN QUANTUM MACHINE LEARNING

Inherited and obscured biases in quantum machine learning models and training data

Like classical systems, quantum machine learning carries risks of perpetuating harmful societal biases embedded in training data or arising from narrowly designed models. In fact, the complexity of many quantum learning algorithms may make recognizing and tracing biases even more challenging.

Without proactive efforts, quantum machine learning could end up unfairly disadvantaging certain groups despite its great promise. Avoiding this failure mode requires careful consideration of ethics and inclusion in developing quantum AI.

Testing and standards focused on algorithmic fairness in quantum machine learning.

To promote fairness in quantum machine learning, new testing procedures and standards will be important to catch biases before real-world deployment. Testing methodologies must be adapted to account for features and intricacy unique to quantum architectures. Standards bodies have roles in researching quantum algorithmic fairness topics and suggesting best practices. Overall, cultivating cultures of responsible innovation is critical across organizations developing quantum AI, so consideration of ethics is built into systems from the start rather than an afterthought.

TRADE-OFFS BETWEEN UTILITY AND PRIVACY IN QUANTUM APPLICATIONS

Balancing innovation and privacy across quantum techniques

Quantum computing requires carefully balancing often competing priorities between realizing useful applications and preserving individual privacy. For example, quantum machine learning models like quantum annealers may enable great advances in areas like healthcare while threatening privacy by exposing sensitive personal data. Similarly, quantum cryptanalysis could provide security benefits while also diminishing personal protections. Thoughtfully optimizing this trade-off will require both technical ingenuity and engaging diverse perspectives to determine acceptable boundaries.

Adaptive policy frameworks to navigate the shifting frontier between utility and privacy

To successfully balance the advantages and risks across quantum applications, policy frameworks will need flexibility to adapt to ongoing advances. As capabilities evolve, the line between utility and violation will shift. What is acceptable today may not be tomorrow. By fostering collaboration between technologists, ethicists, policymakers, and the public, we may find solutions that permit innovation while upholding core values around human dignity and liberty. The aim should be maximizing social benefit at the shifting frontier of the possible.

FAIRNESS IN QUANTUM CRYPTANALYSIS CAPABILITIES AND USES

Potential for biases and disproportionate impacts in cryptanalysis targets and beneficiaries

As quantum cryptanalysis abilities advance, choices around which cryptosystems to target may raise fairness concerns. For example, prioritizing breaking civilian rather than government systems could disproportionately expand state power over citizens. Cryptanalytic insights might also be preferentially shared among allied nations based on ideology or associations rather than merit. To avoid unfair outcomes, developing quantum cryptanalysis should occur transparently and inclusively with multilateral consultation on risks.

Balancing security benefits with principles of justice in deploying quantum Cryptanalysis

At the same time, the security benefits of quantum cryptanalysis against criminals and adversaries are substantial. Like with any dual-use technology, guidelines and safeguards are needed to steer quantum cryptanalysis ethically to socially advantageous ends. International norms against targeting cryptosystems for solely political rather than security aims could help. So, it could incentivize disclosure of vulnerabilities to vendors when feasible. With wisdom and dialogue, the cryptography community can determine policies upholding both justice and security.

BUILDING PUBLIC TRUST THROUGH QUANTUM ETHICS AND RESPONSIBLE DEVELOPMENT

The role of transparency and ethics in fostering public understanding and support

Realizing the transformative potential of quantum information science requires cultivating and maintaining the public's trust. Practising responsible science means considering the broad societal implications of quantum technologies and keeping the public involved as capabilities advance. Initiatives like ethics boards, pilot studies of quantum applications, and inclusive technical standards development can demonstrate that ethical and social concerns are being earnestly engaged. Transparent public dialogue around quantum computing builds understanding and buy-in.

Quantum education, access, and participation to achieve inclusive development

In addition to transparency, ensuring wide accessibility to quantum technologies and participation in the field helps earn public confidence and benefits. Education initiatives play a key role in demystifying quantum science. Policy and funding that remove barriers to entering quantum computing fields enable diverse voices to shape the technology's trajectory. Prioritizing human development alongside breakthroughs demonstrates quantum computing's aim to empower humanity. An inclusive quantum future is a trusted quantum future.

RESPONSIBLE DATA PRACTICES FOR QUANTUM MACHINE LEARNING

Key elements of ethical data collection, management, and use

Developing responsible data practices must be a priority as quantum machine learning comes to fruition. Vital ethical practices include transparency around provenance and potential biases, implementing privacy protections like de-identification and consent requirements, diversity and inclusion efforts in dataset generation, testing quantum models for fairness, and providing redress mechanisms to address errors or harms. Following such practices helps fulfill the promise of quantum machine learning to benefit people equitably.

Importance of collaboration between developers, policymakers, and advocates

Crafting ethical data policies for quantum machine learning systems will require active collaboration between developers, regulators, and public interest advocates. Technologists bring knowledge of emerging capabilities and constraints. Policymakers contribute expertise in governance frameworks. Advocates provide perspective on potential harms to vulnerable groups. Together, these stakeholders can develop nuanced policies and best practices that help avoid unfair outcomes as quantum AI grows more powerful and ubiquitous.

Regulatory Frameworks and Global Collaboration in Quantum Technologies

Realizing the responsible promise of quantum information science requires developing balanced regulations and encouraging multilateral partnerships between nations and organizations. Well-crafted regulations help manage risks to values like security and privacy without stifling innovation. International technical collaborations allow constructive sharing of ideas and standards to achieve global quantum advancement. As quantum technologies become ubiquitous realities, building governance and cooperation centered on ethics provides the best path to shared prosperity and progress.

POLICYMAKING FOR RESPONSIBLE QUANTUM TECHNOLOGY INDUSTRY GROWTH

Balancing Innovation and Ethical Safeguards in Quantum Computing Regulation

As commercial quantum computing matures, regulators face the challenge of crafting policies that foster responsible industry growth. Striking a balance between innovation and ethical safeguards is crucial for addressing societal risks, particularly in critical areas like infrastructure security and privacy. While reasonable regulations are essential, setting the regulatory bar too high initially may impede innovation and delay public benefits. Policymakers should engage in collaborative efforts with industry and researchers to develop adaptive rules that facilitate quantum progress while safeguarding core values.

Fostering Quantum Workforce Development and Economic Growth

In addition to managing risks, governance also means fostering quantum technology's positive potential through funding research, developing skilled workforces, and removing policy barriers to technology commercialization. Strategic investments in quantum education and R&D, especially in concert with industry, can prime economic growth. Creatively adapting export and procurement frameworks where necessary can accelerate global quantum integration. With wisdom, policy can unlock quantum prosperity.

MULTILATERAL FRAMEWORKS FOR QUANTUM TECHNOLOGY COOPERATION

Building International Norms and Agreements around Responsible Quantum Development

No nation alone can address all the complex impacts of emerging quantum technologies. International cooperation is essential to develop shared norms, policies, and technical standards governing quantum capabilities. Diplomatic agreements and multilateral accords can establish boundaries and incentives to prevent destabilizing arms races and reinforce ethical progress. Quantum bodies within broader forums like the UN also have the potential to broker consensus. Overall, global issues demand global perspectives and unity.

Technical Standards Collaboration and Knowledge Sharing:

In addition to diplomatic channels, collaborative development of quantum technical standards and mutually beneficial knowledge sharing are constructive. Platforms allowing leading experts across nations to contribute insights on challenges like post-quantum cryptography in a precompetitive environment enrich collective understanding. Multilateral standardization also smooths global integration of quantum-safe communication networks and other infrastructure. Technical cooperation is thus integral to global advancement.

AVOIDING A DESTABILIZING QUANTUM COMPUTING ARMS RACE

The Risks of Unconstrained Military Uses of Quantum Computing

A key aim of international governance is steering quantum military applications away from destabilizing trajectories. Without constraints, militarization could spark ruinous arms races. Instead of shared security, quantum capabilities could become zero-sum strategic advantages. Cyber warfare risks could also scale new heights. To keep emerging capabilities applied toward stability rather than volatility requires wisdom and a far-sighted perspective across leading nations.

Multilateral Incentives and Norms Against Weaponization and Unchecked Proliferation

Practically, maintaining peace and security while still realizing benefits from defense-related quantum research will require incentives and international norms against weaponization and proliferation. Verification and early warning measures could also identify concerning activities before capabilities spiral out of control. But centrally, a shared ethos that security comes from cooperation rather than dominance is essential.

QUANTUM CYBERSECURITY STANDARDS AND BEST PRACTICES

Securing Quantum Computing Systems and Networks Against Cyber Threats

As quantum computing systems become ubiquitous, cybersecurity emerges as a critical governance priority. Malicious intrusions could steal valuable data or sabotage quantum capabilities. Frameworks promoting resilience against cyber threats should involve developing standards around software assurance, hardware integrity, and continuity of operations. Aligned policies reinforcing strong technical security practices can compel widespread adoption.

Workforce Development and Education for Quantum Cybersecurity

In addition to technical standards, building skilled workforces is also key to quantum cyber readiness. Education pipelines from schools to industry are needed to grow quantum-specific security talent. Centers fostering research on emerging quantum attack vectors and defense strategies can drive innovation. Ample opportunities to enter the field remove barriers and enable idea diversity. Strong teams underpin strong quantum security.

ALGORITHMIC FAIRNESS STANDARDS, TESTING, AND CONTROLS

Technical Standards for Evaluating Bias and Fairness in Quantum Machine Learning

To mitigate discriminatory harms from quantum algorithms, the development of fairness standards, testing procedures, and control mechanisms will be important. Standards bodies have roles in researching quantum-specific metrics and methodologies to evaluate model fairness. They can provide guidance to organizations on requirements for algorithmic audits before deployment. Overall, the widespread adoption of evaluation protocols and controls helps safeguard society.

Policy Frameworks to Incentivize Equitable Quantum Algorithm Development

In addition to technical measures, policymakers have tools to incentivize ethical practices in quantum algorithm development. Potential policy approaches include requiring bias testing for certain applications, mandating transparency around training data and processes, and instituting penalties for deploying discriminatory models. Carefully crafted policy in dialogue with researchers and developers can encourage accountability without limiting innovation.

PROMOTING ACCESSIBILITY AND INCLUSION IN THE QUANTUM WORKFORCE

Improving Quantum Technology Education and Opportunities for All Groups

To fully harness the advantages of quantum computing, education and development opportunities must be accessible to learners and innovators from diverse backgrounds. Policymakers can support community college programs, scholarships, and enrichment initiatives promoting quantum science. Industry leaders can reshape hiring and retention practices to cultivate diverse and inclusive workplaces. Together, these efforts broaden perspectives and maximize idea generation critical to quantum progress.

Partnerships Between Institutions to Widen Access Globally

In addition to national initiatives, international educational partnerships offer avenues to share quantum knowledge and opportunities globally. Exchange programs for students and researchers between universities worldwide foster relationships and skills transfer. Capacity-building assistance can empower developing nations to cultivate indigenous quantum workforces and infrastructure. Accessible quantum education acts as a rising tide that lifts all boats.

INCENTIVIZING COLLABORATIVE AND OPEN-SOURCE QUANTUM DEVELOPMENT

Shared Platforms and Open-Source Ecosystems to Accelerate Collective Innovation

Governance should promote collaborative quantum computing development through shared research platforms and robust open-source ecosystems. Open models empower smaller organizations to build on leading ideas. Widely accessible developer tools and cloud-based services democratize access to quantum power. Transparent, open-source quantum software development allows collective debugging and security. Overall, cooperation propels progress faster.

Grants, Public-Private Partnerships, and Consortia to Fund Open Quantum Projects

Policymakers possess tangible tools to catalyze open and collaborative quantum advancement. Grant programs can fund open quantum hardware and software projects, prioritizing collective benefit. Public-private technology partnerships can facilitate the development of shared platforms, and industry consortia enable firms to pool insights on common challenges and standards. Fostering environments of openness and teamwork leads to shared quantum success.

NAVIGATING TRADE-OFFS IN QUANTUM REGULATION

Balancing computing power concentration risks with centralization benefits

Quantum computing regulation requires carefully weighing trade-offs. While decentralization provides redundancy, some applications like drug development may benefit from concentrating massive computing power. Tighter controls reduce misuse risks but may curb innovation. Blanket access risks abuses, but exclusion based on subjective characterizations is problematic. Navigating these tensions calls for constructive dialogue to find reasonable balance points.

Adapting regulations and oversight to evolving capabilities and risks

Quantum regulatory regimes also need in-built adaptability to technology's rapid evolution. As capabilities advance in areas like cryptanalysis and quantum AI, acceptable boundaries will shift. Finding the right oversight and control touch without needless impedance requires a continuously evolving perspective. Close communication channels between developers, policymakers, and the public enable agile governance to keep pace with technology.

INCENTIVIZING RESPONSIBLE COMMERCIAL DEVELOPMENT

Grants, tax credits, and public-private partnerships for technology companies

Thoughtful policy approaches can steer commercial quantum computing toward ethical outcomes while enabling robust growth. Government grants and R&D tax credits can spur development in priority areas like cybersecurity and medicine over areas with questionable social value. Responsible public-private partnerships allow for aligning innovation to public interests. Overall, good governance provides carrots, not just sticks.

Reasonable oversight and regulations to reinforce responsible industry practices

Of course, prudent regulations and oversight help guard against harmful applications by profit-driven firms. Transparency requirements around capabilities, security controls to protect sensitive data, prohibitions on dangerously autonomous systems, and algorithmic bias testing rules are constructive constraints. Trust develops through dialogue on sensible governance, enabling commercial quantum advancement.

PROMOTING PUBLIC ENGAGEMENT AND MULTIDISCIPLINARY PERSPECTIVES

Fostering public quantum understanding and participation in policymaking

Good quantum governance requires substantive public involvement to represent diverse needs and values. Quantum education, panel participation, citizen juries, and open regulatory comment periods engage the public productively. Accessible legal and funding channels for public interest quantum groups grow participation. Overall, policies should elevate marginalized voices and catalyze wide quantum literacy.

Incorporating multidisciplinary expertise in quantum decision-making

Additionally, effective quantum oversight demands multidisciplinary input spanning technology, security, ethics, law, social science, and beyond. Solely technical perspectives need to be revised to navigate complex societal implications. Multistakeholder quantum advisory bodies, hackathons mixing skill sets, and participatory consensus conferences help broaden decision-making. Blending wide expertise enlightens holistic governance.

CHAPTER 10

FUTURE OF QUANTUM COMPUTING

Quantum Cloud Computing: Quantum-as-a-Service (QaaS) and Cloud Access

Quantum cloud computing refers to providing access to quantum computers over the internet through cloud platforms. This allows users and organizations to leverage the power of quantum computing without needing to buy and maintain expensive quantum hardware and infrastructure. The ability to access quantum computers via the cloud is expected to greatly accelerate the adoption and application of quantum computing across many industries.

QUANTUM COMPUTING CLOUD PLATFORMS

Quantum computing is rapidly moving to the cloud. Leading tech companies like Amazon, Microsoft, IBM, and Google are investing heavily in developing quantum cloud platforms. These platforms allow users to access quantum processors and program experiments and algorithms through online services and APIs. Quantum computing power that was once only available to major research labs can now be accessed on-demand via the cloud.

Key players in the emerging quantum cloud computing space include IBM with its IBM Quantum Cloud service, Amazon with Amazon Bracket, Microsoft with Azure Quantum, D-Wave with its Leap quantum cloud service, Rigetti Computing's Quantum Cloud Services, and IonQ with its IonQ Quantum Cloud API. These platforms provide gate-based universal quantum computers as well as annealing quantum computers optimized for quantum simulation and optimization tasks.

As quantum hardware continues to advance, these quantum cloud platforms are expected to expand and enable more complex applications deployed in hybrid quantum-classical environments. The availability of quantum computing power through cloud access will help drive innovation and adoption across industries.

VIRTUAL QUANTUM LABS AND DEVELOPMENT ENVIRONMENTS

In addition to physical access to quantum processors, cloud platforms are also providing virtual quantum labs and development environments. Platforms like Amazon Braket and Azure Quantum allow developers to create virtual quantum workbenches to design algorithms, test programs on quantum simulators, and integrate with developer tools and workflows.

IBM's quantum cloud service includes Qiskit, an open-source quantum development framework with tools for creating and executing quantum programs. D-Wave offers virtual workshops for coding and running experiments on its quantum annealing systems. Startups like Zapata Computing also provide cloud-based development environments for quantum applications.

These virtual labs enable coders and researchers to build skills in quantum programming and experiment with quantum applications right from their browsers without specialized hardware. The cloud-based dev environments help lower the barriers to entry and promote wider skill-building to create a quantum-ready workforce.

QUANTUM COMPUTING AS A CLOUD SERVICE (QAAS)

Quantum Computing-as-a-Service, or QaaS, refers to the on-demand delivery of quantum computing power and resources via the cloud. Users can access quantum processors and algorithm libraries through QaaS providers on a pay-as-you-go basis, similar to classical cloud computing services. This provides flexible access to quantum computing power without large upfront investments into dedicated in-house infrastructure.

QaaS allows organizations to get started with quantum computing and build expertise at lower cost and risk. With QaaS, companies only pay for the quantum computing time they use rather than multi-million dollar systems. Top players in QaaS currently include Amazon Bracket, Azure Quantum, Rigetti Cloud Services, and IBM Quantum Cloud. As quantum hardware matures, QaaS is expected to spur innovation and accelerate real-world implementation of quantum computing across sectors.

HYBRID QUANTUM-CLASSICAL CLOUD WORKFLOWS

A key trend as quantum computing moves to the cloud is integrating quantum resources into hybrid quantum-classical workflows. Solving real-world problems will require synergistically orchestrating quantum and classical computing. Cloud platforms are enabling these hybrid workflows.

For example, a machine learning pipeline might use quantum algorithms for optimization steps while training the model on classical hardware in the cloud. Quantum annealing could find optimal solutions fed into a larger classical workflow for scenario modelling and analysis. Cloud services that allow smooth integration of quantum subroutines into classical programs and provide quantum simulators for development will be critical for practical hybrid algorithms.

Microsoft, Amazon, IBM, and startups like Zapata are building out the software and infrastructure to seamlessly connect quantum and classical resources in the cloud. As quantum processors advance, orchestrating these hybrid workflows will open the door to tackling a wide range of computationally intensive problems.

EDGE QUANTUM COMPUTING

Edge quantum computing refers to running quantum algorithms and small processors outside centralized data centres closer to where data is generated and collected. This is needed for several reasons. First, today's quantum computers require very precise operating conditions, including extreme cryogenic cooling, which is easier to provide in specialized controlled environments. Second, quantum processing works best on localized data sets before aggregating results to avoid data transfer bottlenecks.

Finally, edge quantum deployments are useful for latency-critical applications like real-time hybrid quantum-classical decision-making. Major players are developing miniaturized quantum modules and sensors that can integrate with edge networks and gateway devices. Startup ColdQuanta is a leader in edge-ready quantum hardware. Their Cold Atom Quantum Technology is engineered for robustness and easy integration beyond traditional lab environments. As quantum computers shrink in size, more processing tasks can be distributed across edge devices and centralized data centres depending on use case requirements.

CLOUD-BASED QUANTUM MACHINE LEARNING

Quantum machine learning, or QML, refers to quantum algorithms specialized for machine learning tasks like classification, regression, and clustering. For certain problem classes, QML promises exponential speedups compared to classical machine learning. Key applications include pattern recognition, fraud detection, traffic prediction, financial analysis, and more. Cloud access will be critical for applying QML techniques to real-world big data.

QML algorithms in development include quantum neural networks, quantum Boltzmann machines, quantum classifiers, and quantum Helmholtz machines. Hybrid strategies combine quantum and classical processing for training and inference. Startups like QC Ware and Zapata are commercializing QML software stacks. Major cloud providers are integrating QML into quantum cloud services. As quantum processors improve, running QML in the cloud will enable tackling compute-intensive AI workloads across industries.

QUANTUM CLOUD SECURITY

Realizing the benefits of cloud-based quantum computing requires securing these services against data theft and tampering. Quantum key distribution via photons enables perfectly secure communications between users and the cloud. Quantum encryption using the one-time pad scheme provides unbreakable ciphertext. Quantum hardware vendors are also exploring qubits on demand for authentication and access control.

Startups like ISARA are developing quantum-safe cryptography tailored to cloud environments. This includes quantum-resistant authentication and encryption protocols to future-proof classical security infrastructure as quantum computers evolve. Industry groups like the Quantum-Safe Security Working Group guide migration strategies. Adopting quantum-safe security techniques will ensure the resilience of the quantum cloud ecosystem.

COST MODEL EVOLUTION

The cost model for quantum computing services will likely evolve as both the technology and business models mature. In early QaaS offerings, basic pricing is via a per hourly rate billed for quantum processor access time. But as algorithms grow more complex, more advanced pricing models will emerge. IBM has proposed metrics like Quantum Volume and T-depth to charge based on qubit connectivity and gate sequence complexity.

To maximize utilization of limited quantum resources, providers may also implement shared availability of oversubscribed systems across users. This is similar to how classical cloud infrastructure is managed, leveraging economies of scale. Dynamic pricing models and shared quantum resource pools will help balance supply and demand cost-effectively as QaaS scales commercially.

CLOUD INTEROPERABILITY AND ALLIANCES

As the quantum cloud ecosystem expands, interoperability standards will be needed to enable hybrid algorithms and workloads that span quantum systems from different providers. Groups like the Quantum Interconnect Alliance are already working to define frameworks for interoperability between quantum cloud platforms and services.

Seamless integration will allow developers to select the best mix of quantum and classical resources across public clouds and optimize complex workflows. Alliances to enable easier migration of data and jobs between quantum providers also promote innovation and avoid vendor lock-in. Shared interfaces and protocols will grow the quantum cloud market collectively.

VERTICAL APPLICATIONS AND PARTNERSHIPS

Leading quantum cloud providers are forging partnerships with major companies across sectors like finance, energy, chemicals, and aerospace to develop vertical industry applications. For instance, JPMorgan Chase and IBM are exploring quantum algorithms for portfolio optimization and risk analysis. Volkswagen Group is leveraging quantum computing for traffic optimization and battery chemistry.

These co-innovation partnerships allow enterprises to tap into quantum experts while providers gain domain knowledge and real business use cases to hone quantum services. Tailored case studies and success stories will help demonstrate ROI and compel more industries to test quantum computing to solve long-standing intractable problems.

DEVELOPER TOOLS AND COMMUNITIES

Comprehensive developer tools and engaged user communities will be critical to accelerate the adoption of quantum cloud services. Platforms need to provide software development kits, emulators, visualization dashboards, and monitoring capabilities. IBM's open-source Qiskit framework is a leading example for building out robust dev tools.

Conferences like Q2B focused on quantum business and hackathons like Q-Hack bringing coders together will grow skills and engagement. Startups like QC Ware also host annual developer conferences centred on quantum software education. User groups foster knowledge exchange and best practices. Vibrant developer ecosystems will feed breakthroughs in cloud-based quantum applications.

CLOUD-BASED QUANTUM WORKFORCE DEVELOPMENT

Quantum computing in the cloud can significantly expand access to education and training resources for professionals to gain in-demand skills. IBM Quantum and AWS Bracket offer extensive learning materials on practical aspects, from quantum basics to coding and algorithm design. Startups like QMinds and QunaSys provide cloud-hosted quantum computing courses for coders.

Hands-on learning with real quantum hardware via simulators makes cloud-based labs ideal for workforce development. Cloud access allows scaling up-skilling programs across organizations. As quantum matures, platforms must continue providing robust training content to grow a quantum-aware workforce able to apply the technology.

THE OUTLOOK FOR QAAS

The Quantum Computing as a Service market is projected to grow rapidly as hardware improves and real-world use cases emerge. Gartner forecasts QaaS revenues could exceed $300 million by 2023. Meanwhile, the cost of entry is falling with cloud access, spurring experimentation. Continued investment in software stacks and expanding cloud ecosystems will drive adoption across sectors in the years ahead. Seamless access to quantum capabilities via the cloud promises to catalyze innovation across industries.

Quantum Internet: Building Secure Quantum Networks

The quantum internet refers to using quantum mechanical effects like entanglement and teleportation to build more secure and faster networks for transmitting information. While still theoretical, quantum communication has made major advances and could provide the backbone for extremely secure networks for critical infrastructure in the future. Realizing the quantum internet will require overcoming key technical challenges.

QUANTUM TELEPORTATION AND CRYPTOGRAPHY

Quantum internet relies on techniques like quantum teleportation and quantum cryptography (QECCs) for transmitting information in ways that are faster, more secure, and have advantages over classical networks. Quantum teleportation uses entanglement to transfer the state of a qubit over distances without having to physically transit the qubit itself. This provides a means for robust qubit transmission over fibre networks.

Quantum cryptography utilizes the laws of physics to create virtually unbreakable encryption keys. Quantum key distribution (QKD) encodes information in photons in a way that would be altered if eavesdropped, creating an ultra-secure channel. Together, teleportation, QECCs, and QKD provide the mechanisms for building extremely secure quantum communication networks.

QUANTUM REPEATERS: EXTENDING ENTANGLEMENT OVER LONG DISTANCES

A key technical obstacle to realizing global quantum networks is extending entanglement over long distances. Entangled qubits easily lose correlation over hundreds of miles of fibre optic networks or free space channels. Quantum repeaters act as relay stations, receiving and re-transmitting quantum information while preserving entanglement between end nodes.

Several approaches to quantum repeaters are being explored, using quantum error correction codes along with matter qubits that can store entangled states for longer times. Startups like QuEra Computing are developing quantum repeater designs based on neutral atoms trapped in optical lattices. Overcoming distance limitations will be critical for any scalable, real-world quantum internet.

PILOT PROJECTS AND TESTBEDS

Pilot quantum network projects have already demonstrated the core components needed for a quantum internet. In 2020, DOE researchers established quantum entanglement over 44 miles of fibre between Livermore, CA, and Sacramento. The Quantum Internet Alliance, led by the University of Chicago, also operates a 20-mile quantum testbed.

The EU Quantum Internet Alliance is developing a quantum network across several European cities, while China recently activated a satellite-to-ground quantum network between Shanghai and Beijing. These testbeds are validating foundational quantum communication protocols and bringing us closer to operational quantum networks.

APPLICATIONS: CRITICAL INFRASTRUCTURE AND FINANCIAL SYSTEMS

Ultra-secure quantum networks could provide virtually unhackable foundations for critical systems and infrastructure. Use cases include connecting military sites or government facilities handling sensitive data. Quantum networks would protect power grids, energy infrastructure, and financial networks from cyberattacks.

Banks are already experimenting with quantum key distribution over fibre optics for encrypting transactions. ETSI is leading the development of QKD standards to secure communications for stock exchanges and central banks. Energy utilities are also trialling QKD point-to-point links. These early applications indicate the transformative potential of quantum networks for securing critical data flows.

REALIZING THE FULL QUANTUM INTERNET

The complete vision of a quantum internet entails integrating quantum routers, repeaters, memories, switches, and networks-on-chip to build a meshed web of quantum connections spanning the globe. This would enable quantum computing in the cloud, connecting data centres housing quantum processors. Scientists also envision native quantum protocols and algorithms that are optimized for the unique properties of quantum networks.

While technical barriers remain, rapid progress is being made, and many experts estimate a functional quantum internet could emerge by 2030. Collaboration between academia, industry, and national labs, alongside continued research and development, will accelerate this vision. The quantum internet promises a new era of communications that is exponentially faster and more secure and could transform our critical information infrastructure.

LOW-LOSS FIBER NETWORKS

One of the biggest challenges for quantum communication is photon loss over fibre networks. Photonic qubits are fragile and decohere easily over long distances. Engineers are optimizing fibre materials, connectors, and couplers to minimize loss. Ultra-low loss fibres below 0.2dB/km have been demonstrated.

Ongoing R&D on hollow-core photonic crystal fibres offers low latency channels for entanglement distribution. Startups like Qunnect are commercializing optimized fibre components tailored for quantum networks. Robust, low-loss fiber optic infrastructure will provide the backbone for long-haul quantum communication over terrestrial networks.

SATELLITE-BASED QUANTUM COMMUNICATION

Satellites can enable global-scale quantum communication by transmitting quantum keys and entangled photons between ground stations. In 2017, China launched the Micius satellite to achieve intercontinental quantum key distribution. The QUESS-2 mission plans to expand this into a continental quantum network.

The European Quantum Communication Infrastructure (EuroQCI) project also aims to build a quantum communication network across Europe using quantum satellite links coordinated through ground-based nodes. Space-based quantum communication will help connect quantum networks across regions efficiently.

QUANTUM NETWORK SWITCHING AND ROUTING

To build out full-fledged quantum internets, sophisticated switches and routers are needed to transmit qubits and distribute entanglement while minimizing errors across complex topologies. Groups like Raytheon BBN are developing quantum switches that use entangled photons for control operations.

Startups like QuEra Computing are also revolutionizing quantum repeater architectures optimized for entanglement routing and pathfinding. Robust switches and routers that can interlink segments and purify quantum signals will be critical network components.

QUANTUM MEMORY NODES

Network nodes with quantum memory capabilities will allow the storing of qubit states reliably during transmission. This is key to enabling longer-range quantum communication. Quantum memories based on trapped ions, defect centres in diamonds, and dopants in silicon are being researched.

ColdQuanta is developing a Quantum Core module that integrates neutral atom qubits with classical controls for managing entanglement distribution and storage. High-fidelity quantum memories with low latency will be integral building blocks of quantum networks.

QUANTUM NETWORK CYBERSECURITY

Robust cybersecurity is critical for enabling trusted communications over quantum networks. Techniques like quantum key distribution already provide hardened communications. Future quantum internet designs must incorporate intrusion detection, secure identity management, and access controls from the ground up.

Startups like Quantum XChange offer solutions like Phio Trusted Xchange for quantum-safe authentication on QKD networks. Optimal cybersecurity solutions must balance rigour with usability for adopters. Security frameworks adopted early can help maintain resilience as quantum networks scale.

INTEGRATION WITH CLASSICAL NETWORKS

The quantum internet will need to interoperate with existing classical networks and infrastructure. Hybrid architectures allow maximizing investments in current fibre and wireless networks while adding quantum-secure communications. Standards bodies are already developing frameworks for combined quantum-classical networking.

Approaches like putting dedicated quantum channels on existing fibre and using software-defined networking for orchestration enable hybrid models. Cisco and other network vendors are working on routers and switches to bridge quantum and classical networks. Convergence will drive adoption during the transition to full quantum networks.

QUANTUM NETWORK MANAGEMENT

Operating and managing quantum networks will require new tools and skill sets combining physics and computer science. Monitoring qubit performance, benchmarking testbeds, and troubleshooting hardware will be critical for maintaining high-quality entanglement links.

Automated control systems will help manage large-scale quantum networks reliably. Quantum system-on-a-chip designs from startups like Quantum Machines provide prototyping environments for robust control software. Advances in quantum network automation & management will enable real-world deployments.

DEPLOYMENTS IN FINANCE AND GOVERNMENT

Early target areas for quantum network deployment are finance and government/military applications where security is paramount. Major banks are already running quantum encryption trials to secure transactions. The EU is planning a quantum communication backbone connecting government centres.

The US and China are accelerating quantum internet R&D to strengthen critical infrastructure resilience. Robust infrastructure for sharing classified data and communications could drive initial rollouts. High-value finance use cases will help demonstrate ROI as quantum networks scale up.

PUBLIC KEY QUANTUM MONEY

Quantum money is a proposed digital currency technology made possible by quantum networks. It uses quantum cryptographic signing of serial numbers on banknotes to prevent counterfeiting and verify transactions. The keys are split between linked nodes in the quantum network.

Startups like Quantum Money Initiative are doing foundational research to make quantum money practical. Central bank digital currencies and payment systems may eventually adopt quantum money for its security benefits. This could enable more secure e-commerce and digital economies.

THE QUANTUM IOT

The convergence of quantum networks with classical Internet of Things infrastructure could enable new paradigms like the quantum IoT. Entanglement could link sensors across vast distances to distributed quantum computers for real-time analytics. QKD provides inherent protection against the hacking of IoT devices.

Quantum radar and LiDAR could enable new sensing capabilities. Automotive, smart spaces, and industrial IoT applications will benefit from the security and performance of quantum communication co-existing with classical wireless connectivity.

OUTLOOK FOR COMMERCIALIZATION

While still emerging, quantum network infrastructure is seeing rising investment and commercialization efforts. Governments are accelerating R&D and testbed initiatives like the European Quantum Communication Infrastructure. Telecom providers like BT, AT&T, and Verizon are ramping up quantum research and trials.

Startups have also raised significant funding, with QuEra Computing securing a $17 million seed round in 2022. Analysts predict global quantum technology revenues to reach $75 billion by 2030, with communications a major segment. Working jointly, government and private sectors can help accelerate this revolutionary evolution in communication networks.

Speculations on Quantum Computing's Long-term Impact: Industries and Society

Overall, quantum computing has the potential to fundamentally transform entire industries and aspects of society. The exponential speedup promised by quantum computers could one day revolutionize complex modelling, artificial intelligence, cryptography, drug discovery, and much more. As this disruptive technology matures, it may reshape finance, energy, healthcare, transportation, national security, and beyond. However, the future remains speculative. This section will explore some possibilities on how quantum computing could transform industries and society decades down the line.

DISRUPTION OF CRYPTOGRAPHY AND CYBERSECURITY

The deployment of quantum computers on a wide scale holds the potential to revolutionize cryptography, which safeguards a spectrum of data ranging from financial transactions to military communications. Quantum algorithms, exemplified by Shor's algorithm, can efficiently compromise established cryptographic methods such as RSA, elliptic curve cryptography, and public key schemes. Consequently, the development and standardization of new quantum-safe encryption protocols become imperative to fortify and re-secure systems. Post-quantum cryptography emerges as a primary focal point in the cybersecurity domain.

Quantum sensors and communication stand poised to bring about a paradigm shift in cybersecurity. Quantum random number generators present the promise of generating unbreakable keys. Quantum key distribution offers channels for data transmission that are perfectly secure. The potential for quantum monitoring of systems introduces the capability to detect tampering, thereby establishing quantum cybersecurity as a foundation fundamentally more robust for IT infrastructure.

FINANCIAL SERVICES AND BLOCKCHAIN

Quantum computing is anticipated to disrupt the financial landscape, impacting domains from risk analysis to trading. Quantitative analysts are already exploring the applications of quantum machine learning for tasks such as stock predictions and portfolio optimization. Quantum simulation holds the potential to expedite risk modelling for insurance and hedging, offering advantages for high-frequency traders.

However, quantum computing also introduces risks to blockchain and cryptocurrencies secured by traditional cryptography. Despite these challenges, ongoing efforts focus on developing quantum-resistant blockchains utilizing hash-based and lattice-based schemes. The emergence of sustainable quantum-secure digital currencies and exchanges remains a possibility, contributing to the establishment of a robust financial infrastructure.

ENERGY AND MATERIALS

From clean energy tech to new materials discovery, quantum computing's predictive power could help accelerate solutions to pressing technology challenges. Simulating chemical reactions and interactions at the quantum level can advance new production methods, polymers, catalysts, and more. Quantum optimization improves everything from renewable energy forecasting to power grid management.

In energy, quantum algorithms are being applied to photosynthesis, solar cells, fusion reactions, and battery chemistry. For discovering and designing novel materials, quantum computing complements AI and machine learning techniques. Quantum advances could help drive the next generation of energy breakthroughs.

HEALTHCARE AND DRUG DEVELOPMENT

Quantum computing is poised to revolutionize health fields from biochemistry to genomics. Quantum AI can analyze patient data and diagnostics for personalized medicine. Quantum simulation of protein folding and molecular interactions provides insights for drug discovery.

Startups like XtalPi use quantum-enhanced algorithms to predict chemical properties and screen drug candidates. Quantum machine learning also accelerates research in targeted therapeutics. Overall, quantum computing could significantly expand capabilities in medical research and treatment.

TRANSPORTATION AND LOGISTICS

From autonomous vehicles to supply chains, quantum computing may transform transportation and logistics. Quantum machine learning improves navigation, planning, and decision-making for self-driving cars and drones. Quantum optimization helps schedule routes, staff, and inventory for complex logistics.

Quantum AI and simulation models can optimize traffic flow in smart cities, predict maintenance for vehicles, and configure efficient warehouses. As systems grow more complex, quantum advantages in optimization, sampling, and machine learning will be essential for managing transportation systems.

SPECULATION ON BROADER SOCIETAL IMPACT

Looking decades ahead, the advent of large-scale quantum computing alongside advances in biotech, AI, automation, and space exploration will usher in sweeping changes whose ramifications remain speculative. Everything from work and education to industrialization and discovery could be transformed. Societal structures may need to adaptively evolve to develop the potential of these exponentially powerful technologies for human progress.

Realizing this vision while managing risks and disruptions will require thoughtful governance and democratization of these game-changing innovations. Nations and societies that plan proactively around the responsible development of technologies like quantum may have a competitive edge in the future. But with the wise and inclusive advancement of science, quantum computing can help build a better future for humanity.

ARTIFICIAL INTELLIGENCE

Quantum computing will likely accelerate the development of artificial intelligence. Quantum machine learning algorithms offer potential speedups for training deep learning models. Large quantum processors could run powerful AI sooner than classical systems. Quantum neural networks may exhibit new capabilities like contextual adaptation. Over time, quantum-powered AI could revolutionize areas from scientific discovery to autonomous systems.

However, risks also need to be considered. While general quantum AI is still distant, it poses complex ethics and alignment challenges research must tease out. But responsibly co-developed, quantum AI could help address global problems that are too complex for current computing. The interplay between quantum and AI will shape the technological landscape of the future in ways hard to predict.

CLIMATE MODELING AND SCIENCE

Quantum simulations of complex quantum systems can provide insights that are useful for everything from drug design to new materials. In climate science, quantum computers may enhance atmospheric chemistry models, simulate ocean currents and ice sheet dynamics, and improve climate forecasting through machine learning.

Startups like OTI Lumionics and Rahko are targeting quantum solutions for solar energy and carbon capture. Overall, quantum advantages in simulating nature could accelerate the discovery of fresh solutions to pressing science and sustainability challenges. Responsible development balancing benefits and risks will be key.

SPACE EXPLORATION AND AERONAUTICS

NASA and space agencies worldwide are actively researching quantum applications from sensing to communications. Quantum gyroscopes, gravimeters, and clocks could enhance inertial navigation for rockets and spacecraft. Entanglement networks could enable faster deep-space communications. Quantum AI promises to optimize spacecraft design and mission planning.

In the distant future, quantum computing may even facilitate a deeper understanding of relativity, quantum gravity, and the origins of the universe - foundational insights that could transform space exploration. Partnering quantum innovation with responsible governance will ensure these breakthroughs benefit humanity.

INTELLIGENCE AND DEFENSE

Quantum computing has extensive national security implications. Codebreaking abilities can compromise legacy encryption and communications. On the other hand, quantum tech enables fundamentally secure encryption, stealth, and sensing capabilities. A quantum arms race of sorts is underway, with nations aggressively investing in quantum R&D.

Maintaining technological leadership will be key. However, ethical risks like mass surveillance enabled by omnipresent quantum sensors require prudent oversight. Quantum research initiatives need civilian balance to ensure security innovations reflect democratic values. Peaceful international collaboration is also critical to prevent unchecked proliferation.

FOOD AND AGRICULTURE

Boosting food production to feed growing populations is a major challenge for the 21st century. Here, too, quantum advances could help. Quantum simulation facilitates engineering higher-yield crops. Quantum machine learning improves predictive analytics for precision agriculture. Quantum sensors enable better soil nutrient monitoring.

Microsoft's recent acquisition of AgTech startup QuantLR highlights growing interest. However, any quantum agriculture solutions must thoughtfully weigh benefits and ecological impacts. Responsible development demands considering sustainability, biodiversity, and climate change challenges holistically.

QUANTUM WORKFORCE DISRUPTION

The transition to a quantum economy will disrupt workforces and require retraining on a large scale. Professions from data analysts to bankers will need quantum skill sets to stay relevant. Technical fields will need project managers versed in quantum complexities. Interdisciplinary talents bridging quantum physics and computing will be highly sought after.

Proactive policies and public-private partnerships are needed for workforce adaptation. Equitable access to quantum education and jobs must be ensured. Responsible transition frameworks will help societies navigate the labour transformations driven by technological shifts of this magnitude.

EDUCATION TRANSFORMATION

Quantum computing could transform education in radical ways. Quantum simulation enables experiential learning of physics and chemistry at the molecular level. Augmented and virtual reality platforms powered by quantum CPUs generate interactive learning environments surpassing in-person labs. Quantum AI could revolutionize personalization for better educational outcomes.

But, care must be taken to balance technology immersion with human development. Curricula will need constant innovation to develop well-rounded critical thinkers and ethical technologists for the quantum age. Responsible development should optimize education to expand human potential.

GEOPOLITICAL BALANCE OF POWER

The quantum race could disrupt geopolitical balances of power. Quantum computing advantages may translate into economic and military dominance. Already, nations are competing aggressively in quantum R&D and talent acquisition. Breakthroughs could confer disproportionate influence on certain countries. Trade barriers risk bifurcating the quantum ecosystem.

Balancing open innovation ecosystems with responsible governance and non-proliferation will be crucial for stability. Mechanisms like the Quantum Economic Development Consortium model multilateral collaboration. International cooperation and transparency can help democratize quantum in ways that diffuse geopolitical risks.

QUANTUM LEGISLATION AND POLICY

As quantum matures, dedicated policies and legislation will be needed nationally and internationally to steer development responsibly. Laws around cryptography, privacy, patents, and national security need regular updating for the quantum age. The ethical risks of quantum surveillance and technological unemployment also demand thoughtful public policy.

Global forums on quantum cooperation are starting to emerge. For instance, the Munich Quantum Valley convenes policymakers worldwide. Inclusive quantum governance that balances innovation, ethics, and human rights will help maximize benefits for society. Proactive policy-making is critical to ensure the quantum's positive potential.

ECONOMIC AND SOCIAL DISRUPTION

Quantum computing enables solving optimization problems with unprecedented efficiency. This carries the risk of significant economic and labour market disruption. As processes across industries become optimized, jobs could be displaced. With proactive transition policies, quantum advances can improve inequality.

But optimization could also help make markets and systems fairer and more efficient if governance keeps pace. Planned adaptation and continuing education can minimize disruption. Economic modelling incorporating quantum technologies will likely play a key role in prudent policy-making for socio-economic resilience.

REIMAGINING HUMAN-COMPUTER INTERACTION

New paradigms of human-computer interaction may emerge from synergies between quantum computing and augmented reality/virtual reality platforms. Quantum sensors could enable wearables that connect seamlessly to contextual quantum cloud services and immersive quantum VR environments.

Startups like Quantinuum are exploring demonstration quantum systems optimized for audio-visual interfaces. Responsibly implemented, quantum-enabled realities could expand human creativity and potential. However, care must be taken to design experiences that uplift users and minimize risks of disengagement from reality.

RETHINKING ENGINEERING AND MANUFACTURING

Quantum computers' ability to model molecular interactions at scale could revolutionize engineering and manufacturing. Quantum machine learning promises to optimize computer-aided design and industrial automation. Quantum simulations facilitate testing product performance and reliability virtually.

Startups are already applying quantum techniques to battery chemistry, airplane wing design, and more. Tomorrow's engineers and operators will need quantum-centric design skills. But, sustainable manufacturing also requires circular economic thinking to be designed into the engineering process from the outset.

FUNDAMENTAL DISCOVERIES

Looking decades ahead, mastering quantum computing may yield foundational insights beyond practical applications. Quantum gravitational wave astronomy could test theories of black hole information paradoxes and cosmic inflation after the Big Bang. Probes of quantum entanglement could reveal deeper principles of reality.

Pushing boundaries of knowledge through quantum platforms must continue being driven by scientific curiosity and the urge for basic discovery that uplifts humanity. Ethical application of these breakthroughs for social benefit will define our collective progress as enlightened civilizations. Responsible innovation is key.

PREPARING FOR A QUANTUM SOCIETY

The deep societal implications of quantum computing highlight the need for proactive preparation by policymakers, academia, industry, and the public. Education and skill-building will enable society-wide quantum readiness. Ethical road mapping helps align development to human values. Global collaboration is key to equitable access.

With prudent governance, quantum advances can profoundly uplift healthcare, sustainability, space exploration, and quality of life for all people. Visionary perspectives will help quantum's disruptive power be harnessed to write an inspiring new chapter in humanity's future. The journey begins with taking the first steps wisely together.

Conclusion

Our journey through the fascinating landscape of quantum computing has come to an end, but the quantum revolution is only just beginning. In this book, we have covered the fundamentals of quantum theory, the hardware and software of quantum information processing, practical applications, challenges, and limitations, as well as speculations on the future of the field. As we reflect on the ground covered, let us recap some key lessons and look ahead with excitement to the transformative impact quantum technologies can have on industries, scientific discovery, national security, and society.

First and foremost, we have seen that quantum computing represents a fundamentally different paradigm from classical computing. By leveraging unique quantum phenomena like superposition, entanglement, and interference, quantum computers can solve certain problems that are intractable for even the most powerful classical supercomputers. Algorithms like Shor's for factoring large numbers and Grover's for unstructured search demonstrate quantum speedups over the best-known classical approaches. While the theory has been known for decades, the engineering hurdles to build quantum machines are only now being surmounted in laboratories worldwide.

At the hardware level, we have explored various qubit modalities, including superconducting circuits, trapped ions, and exotic topological states of matter. Each approach has relative strengths and weaknesses in scalability, coherence times, and controllability. Meanwhile, software stacks and cloud services are emerging to make programming quantum computers accessible to those with only classical programming experience. Abstraction layers and simulation tools help hide the underlying complexity of quantum circuits. As hardware and software continue to mature, we will reach the long-anticipated milestone of quantum advantage - solving valuable problems that are classically intractable.

Applications of quantum computing already underway today span optimizations in finance, machine learning, chemistry, and logistics. We are just scratching the surface of this technology's potential. Looking further ahead, a fully fault-tolerant quantum computer would render current encryption schemes obsolete overnight. This looming threat motivates designing new cryptosystems resistant to quantum attacks. Profound impacts on cryptography, cybersecurity, and privacy protections lie ahead.

Of course, realizing the full promise of quantum computing remains challenging. Decoherence and noise must be combatted with error correction and other techniques to achieve accurate, reliable computations. The number of physical qubits needed for fault tolerance likely ranges from 500 to millions - a daunting scale-up from the few dozen available today. Significant hardware improvements will be needed to reach this goal. Even with logical error-corrected qubits, the limited connectivity between them on near-term devices presents obstacles to running meaningful quantum algorithms. Continued research and discovery are critical to overcoming these hurdles.

Beyond the technical challenges, we must also grapple with the ethical dilemmas raised by quantum technologies. Quantum computers could exacerbate existing biases in data or disproportionately benefit those with access to this scarce resource. Developing regulations to ensure the responsible use of quantum capabilities will be pivotal. Ethicists, policymakers, and quantum scientists must work closely together to steward the societal adoption of quantum computing.

The coming decades will prove decisive in how quickly quantum computing fulfills its disruptive potential. If current exponential progress persists, we may reach the fault tolerance threshold by 2030 or soon after. How exactly this technology will then transform society remains speculative. Entire industries and business models could shift as new optimization capabilities get deployed. Harmful monopolies built on computational asymmetry may dissolve if access to quantum computing becomes more democratic. Regardless, the brightest quantum minds must partner with stakeholders across public and private sectors to ensure these powerful capabilities benefit humanity.

For now, much work remains to realize the dream of full-scale fault tolerance. But we are immensely fortunate to be witnessing the beginnings of the quantum era. The progress made in just the past few years is astonishing. It is my hope that the concepts and techniques presented in this book provide you with a foundation to engage productively with quantum computing as it continues advancing. Apply your knowledge to push quantum technologies forward or perhaps discover new quantum phenomena waiting to be unleashed. The future pioneers and visionaries of quantum computing could very well be inspired readers of this book.

Our journey here may be concluding, but yours may have just begun. I encourage you to continue expanding your quantum skills and perspectives. Read voraciously, program quantum circuits, attend conferences, and engage with the growing quantum community. Scale unexplored peaks of human knowledge that may one day be surpassed only through quantum capabilities we have yet to conceive. The present quest to realize quantum computing reminds us that our collective understanding of nature's deepest mysteries is still unfinished. Cherish this sense of wonder and possibility.

We stand today on the precipice of a computing revolution unlike any before. The very fabric of our reality permits quantum phenomena with no classical analog. Harnessing these counterintuitive effects opens new frontiers in computational power, communications, cryptography, and sensing. Yet the true limits are still unknown. What new possibilities may emerge as quantum technologies mature? Will quantum computing be to the 21st century what the semiconductor was to the 20th? The only certainty is that the quantum future will be shaped by the tireless efforts of researchers, engineers, programmers, entrepreneurs, and visionaries like you. I am honored to have shared this journey with you through the pages of this book, and wish you the very best in your ongoing pursuit of the quantum frontier.

Made in the USA
Las Vegas, NV
17 May 2024